RISK AND UNCERTAINTY ANA
SUSTAINABLE URBAN WATER SYSTEMS

Risk and Uncertainty Analysis for Sustainable Urban Water Systems

DISSERTATION

Submitted in fulfilment of the requirements of

the Board for Doctorates of Delft University of Technology and of the Academic Board of UNESCO-IHE Institute for Water Education for the Degree of DOCTOR

to be defended in public

on Wednesday, May 1st, 2013, at 10.00 hours

in Delft, The Netherlands

by

Krishna Bahdadur KHATRI

born in Dailekh, Nepal

Master of Science in Integrated Urban Engineering

UNESCO-IHE, the Netherlands

This dissertation has been approved by supervisor(s):

Prof. K. Vairavamoorthy.

Members of the Awarding Committee:

Chairman	Rector Magnificus TU Delft
Vice-Chairman	Rector UNESCO-IHE
Prof. dr. K. Vairavamoorthy	USF/ UNESCO-IHE/TU Delft, Supervisor
Prof. dr. S.T. Khu	University of Surrey, UK
Prof. dr. S. Mohan	IIT, Madras, India
Prof. dr. ir. C. Zevenbergen	TU Delft / UNESCO-IHE
Prof. dr. ir L. Rietvelt	TU Delft
Prof. dr. ir. P. Van der Zaag	TU Delft / UNESCO-IHE (Reserve)

CRC Press/Balkema is an imprint of the Taylor and Francis Group, an informa business

Published by:

CRC Press/Balkema

PO Box 11320, 2301 EH Leiden, The Netherlands

E-mail: Pub.NL@taylorandfrancis.com

www.crcpress.com – www.taylorandfrancis.com

ISBN: 978-1-138-00096-4

ACKNOWLEDGMENT

This work has been realized with the support and encouragement of several people to whom I am very grateful. First, I would like to express my most sincere thanks to Prof. Kala Vairavamoorthy, Dean of the Patel College of Global Sustainability (PCGS) at the University of South Florida (USF) who gave me the opportunity to undertake this work and for tirelessly guided me throughout my study. I am extremely grateful for his continuous encouragement, critical observation, comments and valuable time. I thank him hugely for allowing me to complete my study by inviting me to join his research group at the PCGS at the USF and providing me financial support. Prof. Vairavamoorthy was very helpful to me not only in my studies but also in my private life. His generous support was key in all this endeavour and this work would not have been completed without his dedicated support.

This study was financially supported by Delft Cluster and SWITCH. I am very thankful for the generous support that enabled me to complete my PhD study. This study would never have been realised without this financial support.

I am greatly indebted to the research team and staff of PCGS. Everyone at the school was supportive and provided meaningful contributions to my study. I would like to thank Kebreab Ghebremichael, Jochen Eckhart, Seneshaw Tsegaye, Jotham Sempewo, James Buckingham, Garrick Aden-Buie. This wonderful group provided encouragement and support when I most needed them. They helped me by reading my thesis, sharing their valuable comments, and made my life easy in difficult times. I thank you all so much.

I would like to thank staff, friends and colleagues at UNESCO-IHE for their continuous support and friendship. Thanks to Tanny Van der Kliss and Chantal Groenendijk - you were the contact person in the department at UNESCO-IHE and helped me to translate my PhD thesis preposition in Dutch as well. Many thanks to Jolanda Boots who was mostly contacted at IHE and always supporting me since 2004. I would like to thank Anique Alaoui-Karsten, Erick de Jong, Stephanie Petitjean for helping me to process the PhD exam and Peter Stroo for working for printing the thesis. Thanks to Saroj Sharma, Krishna Prasad, Shreedhar Maskey, and Assela Pathirana-you all have advised and supported me by different means. I would like to thank Jan Herman Koster and many others for their help when I needed them. My special thanks to Mariska Ronteltap for helping me the Dutch translation of the summary of this thesis in a very short time. The support and encouragement I got from my friends and PhD colleagues Durga Lal Shrestha, Zhou Yi, Harrison Mutikanga and others are highly appreciated.

I would like to thank the Department of Civil Engineering, University of Birmingham for hosting me to undertake my research. I really appreciate their hospitality and support during my stay in Birmingham, UK for more than 3 years.

I would like to extend my appreciation to the members of the PhD awarding committee for their time, invaluable comments, and suggestions.

Finally, I am extremely thankful to my wonderful parents/in-law and family, who understood the challenges and supported me without any hesitation. I lived far from you for years— while you desperately needed me. You always motivated and suggested me to control my sentiment while the situations were beyond the control - I am deeply indebted. I greatly thank my wife Jeena for being very patient during all these years. This journey was completely unexpected, unplanned and very difficult for you. We moved in three countries and many apartments. You lived far from your family. You manage everything. I am very proud of you, and hugely value your contributions to my study. My lovely daughter, Abhilasha, I thank you so much for being a nice girl and for being patient at all times. I am very proud of your excellent school performances irrespective of changes in this journey from Kathmandu, Nepal to Birmingham, UK to Tampa, USA.

SUMMARY

Long-term future planning is not a new approach in urban water management (UWM). However, the conventional 'stationary approach' of infrastructure planning and decision-making, where the future is assumed as the continuation of historical observation, will not work in the rapidly changing environment. This is because the current and future change pressures, such as climate change, urbanisation, population growth, deterioration of infrastructure systems, and changes in socio-economic conditions are always uncertain. Uncertainty in future change pressures stems from two quite different sources: incomplete knowledge and unknowable knowledge. Incomplete knowledge is due to lack of information and understanding of a system. Unknowable knowledge is due to the inherent indeterminacy of future human societies and both natural and built systems.

Urban water systems (UWS) are complex and dynamic. It comprises of numerous interacting components and multiple decision-makers including users, water companies, municipal boards, and other authorities. Such complex interactions and interdependencies transcend multiple sources of uncertainty during the planning and decision-making process. These uncertain change pressures generate multiple direct and indirect impacts upon UWS. Due to this, it is difficult, and in some cases impossible and impracticable to analyze accurately the magnitude of impacts in a system. The deterministic approach to infrastructure system design can not capture such characteristics of a complex system and uncertainties associated to the future change pressures. As a result, the risk of not meeting the expected level of system performances and not being cost effective is very high. This research discusses the major future change pressures, associated uncertainties, and risks for sustainable infrastructure system planning.

Most of the existing uncertainties modelling techniques are based on either probability theory or fuzzy set theory. Real world problems, however, are associated with the multiple types of uncertainty (i.e., aleatory and epistemic), and in many cases data scarce. The literature review on uncertainty modelling acknowledge that a hybrid approach to uncertainty modelling is required to analyse both types of uncertainty in a single framework. This could be achieved either by (i) transforming all the information to the probabilistic or fuzzy form or propagated homogenously through a model (i.e., either random simulation or fuzzy approach), or (ii) describing all the information separately and propagating heterogeneously. However, there is limited guidance to select an uncertainty analysis technique according to their sources and types. Moreover, most of the existing decision-making frameworks developed for risk-based decision-making or multi-criteria analyses do not address uncertainty appropriately. This thesis develops uncertainty analysis framework and methods that can capture the multiple sources and types of uncertainty in a single system.

The decision-making process requires addressing the conflicting social, economic, and environmental objectives. This can be achieved using a multi-criteria decision analysis (MCDA). The published literature on multi-criteria decision analysis show that the decision results calculated from very simple techniques to complex ones was not significantly different. It was reported that MCDA methods are either not user friendly, not transparent or do not allow for stakeholder engagement. Thus, the need for simple and transparent MCDA method that ensures the engagement of stakeholders was acknowledged. Moreover, uncertainty is unavoidable in measuring and representing the performance criteria. Therefore, the literature acknowledge the need of MCDA method that can capture the uncertainty associated within the performance criteria and technique for aggregation of multiple sources of information that could be a crisp, probabilistic, and qualitative type in a single framework.

This research aims to develop a decision-making framework, methodology, and tools for analysing the risks and uncertainties in UWS. The research was undertaken through theoretical and modelling approach and demonstrations of the developed framework. It has the following specific objectives:

- To develop a risk assessment framework for urban water systems considering future change pressures and their associated uncertainties.
- To develop a hybrid approach to uncertainty analysis to describe and propagate the multiple types of uncertainty in urban water system models.
- To develop a multi-criteria decision-making framework that is tailored for UWS operating under uncertainty.
- To demonstrate the applicability and effectiveness of the developed frameworks and methods by applying them to real-world UWS.

The existing literature defines risk vaguely. It is interpreted as a combination of the likelihood of a negative event (e.g., what can go wrong) in a system and the corresponding consequences. This research defines risk as a violation of the minimum expected level of performance and the associated negative consequences. Thus, risks in UWS are perceived, if the expected performance of the system is violated.

Risk in a system is assessed using either a quantitative or a qualitative method. Qualitative methods use expert opinions to identify and evaluate the probability and consequences of failure. Commonly used qualitative methods include safety reviews, checklists, "what if" scenarios, preliminary hazard analysis, and hazard and operability studies. On the contrary, quantitative methods rely on information databases and statistical methodologies. A few of the commonly used quantitative methods of probabilistic risk analysis include failure modes and effects analysis, fault trees, and event trees. The identified risk in a system is reduced or controlled by employing different measures suggested for risk management. Risk management is the process of selecting a strategy to control or reduce the identified risk. Considering the complexities of UWS, the proposed risk analysis framework employs both the probability and fuzzy set theory to analyse the risk of failure of system performances and MCDA method for selecting the best strategy for the risk management.

This research develops a hybrid method to uncertainty analysis. In this method, all the input parameters of a model is transformed into one form (i.e., either into the probabilistic or fuzzy form) and propagated homogenously through a model. Dempster-Shafer theory is employed to develop transformation algorithms. The transformed uncertain information can be propagated using any well-established technique, such as Monte Carlo Simulation, Latin Hypercube Sampling, Bootstrap Simulation and fuzzy alpha-cut techniques. The output of the analysis is presented in an uncertainty envelope with statistical information and a 95% confidence interval.

A fuzzy set theory based multi-criteria decision-making framework is proposed to address the complexities and multi-disciplinary aspects of UWS. The multi-criteria are selected in different dimensions of infrastructure systems (e.g., effectiveness, reliability, system cost) and multiple objectives of analysis (e.g., environmental, social, economic, etc.). The criteria and their indicators are represented considering the types of information contained, for example, crisp or probability distribution function or fuzzy membership functions. The composite index is computed using the fuzzy synthetic evaluation technique. The relative importance of the indicators is defined by applying an Analytic Hierarchical Process (AHP) based algorithm. The index represents the aggregated performance of a system or the system's components. The index is also used for selecting a suitable option/strategy for risk management.

This research applied the decision-making frameworks, methodologies, and tools in three different cases. The first case analyzes the risk involved in future water availability in Birmingham, UK by 2035. The major future change pressures considered for the assessment include climate change, population growth, socio-economic change, and water losses from the system. The hybrid method to uncertainty analysis is applied in this case study. The second case is performance-based urban infrastructure system planning in Kathmandu, Nepal. It demonstrates the applicability of MCDA method for integrated urban infrastructure decision-making. The third case demonstrates the urban water sources and demand analysis in a limited data case situation in Mbale, Uganda. It shows how risk based decision-making framework can be applied for UWS planning and decision-making.

Apart from the critical literature review on defining the risk and uncertainty analysis concept, frameworks and methods for uncertainty analysis and multi-criteria decision-making, the main contributions of this research are:

- Development of a new risk assessment framework that proposes a hybrid method to uncertainty analysis, risk assessment, and a fuzzy set theory based multi-criteria analysis method for risk management and decision-making.
- Development of a hybrid approach to uncertainty modelling based on evidence theory to analyse the multiple types of uncertain information in a single framework.
- Development of a fuzzy set theory based multi-criteria decision-making framework and technique for performance based decision-making and selecting a sustainable strategy for risk management.
- Application of the developed frameworks, methodologies, and tools for UWS planning and decision-making in developed and developing countries cases.

The new risk assessment framework developed from this study captures different types and sources of uncertainties in a single framework and is applicable in other fields of infrastructure planning without further modification. The hybrid approach to uncertainty analysis is a way forward for appropriately representing and propagating uncertainties in a complex system. The MCDM framework allows selecting a better option/strategy for risk management in UWS. It addresses multi-dimensional issues of infrastructure system planning and the role of stakeholders during the decision-making process. The fuzzy set based multi-criteria framework for decision-making is flexible enough to capture the multiple types of data available in real cases. The demonstration of developed framework, techniques, and tools in real cases of UWS both in developed and in developing countries shows the utility of the research outcomes.

Krishna Bahadur Khatri

UNESCO-IHE, Delft

TABLE OF CONTENTS

Chapter 1   Introduction

This Chapter introduces the research objectives, research scope, and thesis structure. It starts with the introduction of urban water systems (UWS) and its relation with the concept of sustainability. It then provides an overview of major future change pressures and the associated sources of risk and uncertainty in UWS. A current state of knowledge describes the need for further research on risk and uncertainty analysis for sustainable urban water systems. Based on the identified research need, the objectives and research questions are outlined. The structure of the thesis is presented at the end of this Chapter. The Chapter is based on the papers Khatri et al. (2007); Khatri and Vairavamoorthy (2007); Khatri and Vairavamoorthy (2013a).

1.1   Context of the Study

UWS are characterized by an urban water cycle similar to the hydrological cycle located within an urban catchment. UWS are often two mixed cycles: the water and nutrient cycles (Butler and Maksimović, 2001). The water cycle includes the various water streams throughout an urban environment including surface water, ground water, rainwater, and wastewater, whereas the nutrient cycle is related to the flow of nutrients, such as Nitrogen and Phosphorous. UWS are formed through interconnections of water sources, water treatment plants, water conveyance and distribution networks, sewer and drainage systems, wastewater treatment plants, and receiving water bodies. UWS supply water for both industrial production and economic development, they transport waste away from the city, drain storm water, and maintain the integrity of receiving water bodies and urban streams (Figure 1.1). UWS are thus an integral part of urbanized life. The quality and level of services of UWS define the functions and values of a city and determine the living standards and well-being of a city's population.

UWS are complex systems consisting of social, technical, and environmental systems. The social system considers end-users, their behaviours, and the interactions between them. The technical system includes infrastructure and processes such as pipes, ponds, and treatment facilities. The environmental system comprised of the surrounding environment, especially the recipient, the soil, the atmosphere, and certain natural processes (e.g., meteorological and aquatic) that take place in these locations. These three systems are interdependent and interconnected within the three major infrastructure sub-systems of water supply, wastewater, and storm drainage (Khatri and Vairavamoorthy, 2007).

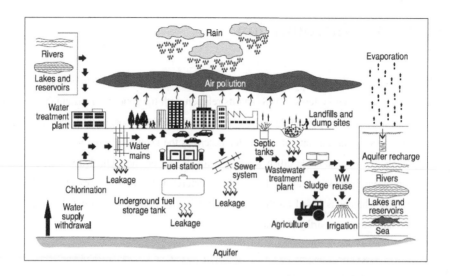

Figure 1.1 The urban water systems (Marsalek et al., 2006)

Over the past decade, sustainable urban water development has become an important development agenda. The Brundtland Commission Report (WCED, 1987) defines sustainable development as development that meets the needs of the present without compromising the ability of future generations to meet their own needs. In this approach, sustainability implies that the supply of 'natural capital' is maintained such that the use of renewable sources–such as water withdrawal–should not exceed the rate of renewal. The use of resources should be such that they will not be exhausted before alternative sources are available, and at the same time fundamental ecological processes and structures should be maintained   (Rijsbermana and Van de Ven, 2000). Main objective of the sustainable development is, therefore, environmental protection should not preclude economic development, and economic development must be ecologically viable now and in the long- run. It requires an integration of economic, social, and environmental polices.

Many researchers in UWS have found that the current practices of urban water management are not sustainable (see, Biswas, 1991; Pinkham, 1999; Bouwer, 2002; Elimelech, 2006; Malmqvist et al., 2006; SWITCH, 2006; Graedel and Allenby, 2010; Farrelly and Brown, 2011). Those studies have shown that the objectives of sustainable UWS must establish efficient and integrated management of water resources by all water-consuming sectors—people, industry, agriculture, and ecosystems—such that the harmony with the bearing capacity of nature can be maintained. Some joint goals across all sectors include the sustainable use of water resources; more efficient water use through conservation and reuse of water; reduction in energy consumption; production of valuable products from wastewater; and a provision of quality water that is fit for all purposes. Yet, those objectives are still unmet in most urban centres and UWS.

Moreover, sustainability necessitates consideration of the needs of future generations. However, the analysis of future needs is always challenging due to dynamic and uncertain future change pressures. Major future change pressures, such as climate change, rapid population growth, evolving urban land use, economic development—may alter water availability, modify water demand, or introduce new pollutants (Khatri and Vairavamoorthy, 2007). The impacts of these pressures are wide-ranging and difficult to predict precisely, this indicates that sustainable UWS necessitates the consideration of uncertainty.

Planning for a sustainable UWS also requires analysis of multiple, yet, conflicting social, economic, and environmental objectives. The analysis has to cover the future perspectives, particularly the associated risks and uncertainties, as the performances of systems will be affected by both the internal and external future change pressures. According to the SWITCH (Sustainable Water Management in the City of the Future) approach, the risk and uncertainty analysis is required to achieve sustainable UWS (SWITCH, 2006). This research aims to identify such decision-making framework for the strategic planning of UWS, particularly focusing on risk and uncertainty analysis that hinder the objectives of sustainability (i.e., minimum level of services to public life and environment).

1.2   Risk and Uncertainty in UWS due to Future Change Pressures

Risk and uncertainty analysis

Risk is often used interchangeably with terms such as chance, likelihood, and probability to indicate uncertainty about the state of items under discussion. Most meanings of risk carry negative connotations of danger, peril, or loss. The particular risk of interest in this study is to identify the undesirable events in UWS due to future change pressures. The term "undesirable events" is defined as a violation of the minimum expected levels of service in a system. Thus, a risk in a water supply system will be perceived if internal or external change pressures have violated the system's performances. In this discussion, system performances refer to the minimum level of services, which might include water availability, water quantity, and quality supplied to the customers, and reliability of the services against failure (further detail see Chapter 2).

Uncertainty is a state of knowledge between certainty, which is perfectly known, and the present state of information, which may be partially known or completely unknown. This stems from the natural variability of systems, the complexity of social interactions in built systems, and complexities in system models and model processing. Uncertainty, thus, describes the quality of knowledge concerning risk that needs to be faithfully represented, propagated, and presented for decision-making. Uncertainty analysis methods depend on the types and sources of uncertainty. The uncertain information may be stochastic, random, vague, or subjective. Sources of uncertainties refer to the location and causes that give rise to an uncertainty (see Chapters 2 and 3 for detail).

Future change pressures and sources of risks and uncertainty in UWS

Managing a sustainable UWS is more challenging than ever, largely due to the risks and uncertainties posed by global change pressures. Global change refers to changes in human life and the environment resulting from the impacts of global dynamics. Price (1989) defines global change as changes in the environment that are human-induced, including climate change, ozone depletion, acidification, land degradation, desertification, and dispersion of chemicals, to name a few. Rotmans et al. (1997) denote global change as the totality of changes evoked by complex mutual human-environment relationships. It specifically addresses geo-physical, biological, chemical, and ecological components and categorizes them into social, economic, and ecological dimensions. Since the focus of this research is to develop a decision-making framework considering future change pressures that influence the management of UWS. Thus, this research includes both global (e.g., climate change) and local (e.g., deterioration of infrastructure systems) as future change pressures.

The key future change pressures most often discussed include: (i) climate change and climate variability, (ii) population growth and urbanization, (iii) deterioration of infrastructure systems, (iv) emerging pollutants, (v) increased risks in critical infrastructure systems, (vi) increasing energy price, (vii) changes in governance and privatization, (viii) emerging technologies and (ix) globalization and economic development (see, Figure 1.2).

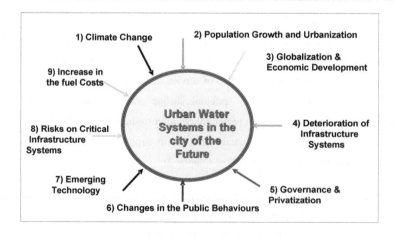

Figure 1.2 Global change drivers for the city of the future (Khatri & Vairavamoorthy, 2007)

Several studies have discussed on many future change pressures in UWS (Rotmans et al., 1997; Kelay et al., 2006; Khatri and Vairavamoorthy, 2007; Segrave, 2007). Considering the magnitude and extent of the impacts of these major change pressures in UWS (Khatri et al., 2007; Mukheibir and Ziervogel, 2007; Ruth et al., 2007; Mehrotra et al., 2009; Khatri and Vairavamoorthy, 2013a), this research focuses on the main four future change pressures: climate change, population growth and urbanization, socio-economic change, and deterioration of urban infrastructure systems as discussed below.

i) Climate change

Climate change is predicted to cause significant changes in precipitation patterns and their variability, which will influence the UWS differently, such as availability of water resources, urban flooding, and increase in emerging pollutants in water sources.

The risk to water availability primarily relates to changes in the hydrological cycle and extreme weather events (Alcamo and Kreileman, 1996; Vörösmarty et al., 2000; Alcamo et al., 2007; IPCC, 2007a). It is expected that a greater proportion of winter precipitation will fall as rain rather than as snow. With the rise in mean temperature, glacier retreat will continue to the point that many small glaciers are likely to disappear. These projected changes in precipitation will lead to variances in the direction and magnitude of groundwater recharge. Additionally, in recent years more regions have been experiencing droughts (IPCC, 2008) from enhanced evapotranspiration and reduced soil moisture brought on by decreased land precipitation and increased temperatures (Hultman et al., 2010). The impacts of droughts could be seasonal, but will ultimately impact water availability. Moreover, sea level rise and salt-water intrusion will have an impact on local water sources availability near coastal areas.

Climate change will affect water quality in various ways (Delpla et al., 2009). The increase in water temperature will alter the operation of bio-geo-chemical processes (degrading and cleaning) and lower the dissolved oxygen concentration of water. Increased occurrences of higher runoff will increase the pollutant loads (suspended matters, NOM, and heavy metals) and overflow sewers. Moreover, increased flooding along with overflow of treated or untreated wastewater sewer systems will cause serious effects on the biotic life cycle and increase possibility of outbreaks of water borne diseases (Schiedek et al., 2007). Water quality issues may be more pronounced in lakes due to higher incidence of the eutrophication process (Hellmuth and Kabat, 2002).

Droughts also reduce water quality due to high water temperatures and low river discharges, which result in limited dilution of the chemical load from point sources. Higher surface water temperatures promote algal blooms and increase bacteria and fungi content (IPCC, 2007b). This may result to odour and taste problems in chlorinated drinking water and the occurrence of toxins. The long term warming will lead to algal growth even with enhanced phosphorus removal in wastewater treatment plants (IPCC, 2007b). Moreover, there is likely rendering a water source unusable unless special treatment is introduced due to increase in nutrients and sediments load and lower water levels.

It has been reported that changes in climate variability and an intensification of the hydrological cycle have led to more frequent flooding (IPCC, 2012; Jha et al., 2012). Some specific contributory climatic factors discussed in the literature are changes in local rainfall patterns that could lead to more frequent or intense river flooding or more unpredictable flash flooding; sea level rise in coastal areas; and increasing frequency of storms that can produce sea surges. The processes that induce flooding include intense or long-lasting precipitation, snowmelt, reduced conveyance due to ice jams or landslides, and storm systems (IPCC, 2007b).

Although numerous risks associated with climate change in UWS are clearly identified, it is difficult to predict their magnitude and intensity due to a series of uncertainties cascading from scenarios down to the impact study in an urban scale (Carter et al., 1999; Allen et al., 2000; Gober et al., 2010; Moss et al., 2010; Wilby and Dessai, 2010; IPCC, 2012). Some of the main uncertainties include: (i) uncertainties associated with the scenarios on how the future as a complex non-linear dynamic system will incorporate all of the emissions generation, natural processes and variability; (ii) uncertainties added by the different global climatic models used for prediction and their capacities; (iii) uncertainties associated with hydrological modelling; (iv) uncertainties associated with downscaling and error; and (v) uncertainties associated with the impact model. Thus, capturing the associated uncertainties due to climate change is still challenging (IPCC, 2012) particularly in places where databases on climate change observations are lacking.

ii)   Population growth and urbanization

Rapid change in population growth and urbanization will have direct impact on UWS through dramatic increase in water demand, wastewater generation, or runoff rates. Cause and effect relations of population growth and urbanisation in UWS is presented in Figure 1.3.

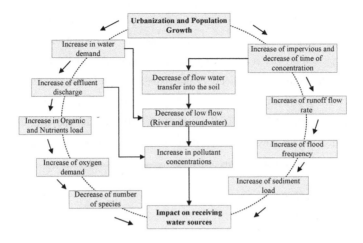

Figure 1.3 Cause and effect relations of urbanisation and population growth in UWS
(Modified after Krajeswski et al., 2000)

Population growth increases water demand and wastewater discharge. Urbanization changes land uses and necessitates the extension of existing infrastructure systems. The extension of paved surfaces will reduce the rate of infiltration and increase the evapotranspiration that leads to depletion of subsurface water, increase in runoff rate and accelerated peak of urban flooding. Moreover, increased urban activities will escalate the atmospheric load of contaminants, such as sulphate, mineral dust, and black carbon aerosols, with the potential to affect the water cycle by suppressing rainfall in polluted areas (Ramanathan et al., 2001). The high concentrations of nitrates, pesticides, heavy metals, hydrocarbons, and chlorinated hydrocarbons will impact the quality of the existing ground water (European Environment Agency, 1999; Roccaro et al., 2005).

The risks of urbanisation and population growth in UWS are well discussed in the literature (Vörösmarty et al., 2010; McDonald et al., 2011). However, there are many sources of uncertainties associated with urbanization, population growth and their interactions with water. For example, the demographic forecasting in an urban area depends on how the rates of fertility, mortality, and migration will change in a region. These factors are affected by many external factors such as war and refugee movement and natural disaster as well as internal factors such as change in demographic policy. On the other hand, the rate of urbanisation is governed by many uncertain pull and push factors. The pull factors include opportunities provided by better social services and securities, employment and better livelihood that attract people to move from rural to urban areas, or towards a larger city area from a smaller town. Push factors, such as a harder rural livelihood, the impact of natural calamities and a lack of physical utilities, services and securities, encourage people to migrate towards cities.

In the literature, probabilistic and scenario approaches are used for demographic forecasting (NRC, 2000). In the scenario approach, a consistent set of assumptions are chosen and embedded to provide a comprehensive picture of what a particular future might look like. Unlike scenarios, probabilistic approaches account for uncertainty explicitly by projecting trends of fertility, mortality, and migration and deriving a resulting probability distribution for the projected population sizes and age structures (Lee and Tuljapurkar, 1994; NRC, 2000; Keilman et al., 2002; Wilson and Bell, 2004). In reality, most of the planning analysis, however, is undertaken without considering those uncertain factors, and in many cases data are not available for the analysis.

iii) Deterioration of infrastructure systems

Underground pipe infrastructure systems are ageing and deteriorating at a faster rate. This creates technological and financial challenges in UWS in regard to maintaining and upgrading infrastructure assets

The failure mechanism for underground pipelines is best described by the "Bathtub Curve" (Figure 1.4) (Klutke et al., 2003). As shown, water pipes follow a continuous deterioration process characterized by three general phases. In the first phase, failure occurs during the early life of pipe installation and is generally caused by errors in design, manufacturing, or assembly. The second phase is a random failure of a relatively constant but low rate that can be observed after the pipeline components are settled. Finally, the third phase usually begins after some years of operation and accumulating damages. Once this phase begins, the deterioration rate increases exponentially until the pipe fails (see further, (Sægrov et al., 1999; Kleiner and Rajani, 2001; Ana and Bauwens, 2010)

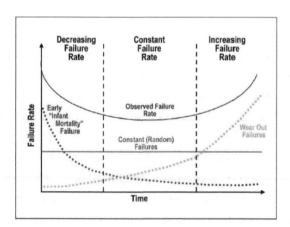

Figure 1.4 Typical Bathtub curve for water pipelines deterioration (Klutke et al., 2003)

As the assets get age, pipes may exposed, and the deterioration process begins. The deterioration rate is related to the various characteristics of pipes and surroundings including location and soil types, characteristics of the liquid flowing through the pipes, types of pipe materials, period of construction, diameter and gradient of the pipelines. As the pipes age, service reliability reduces gradually until it fails completely if they are not rehabilitated (Figure 1.5).

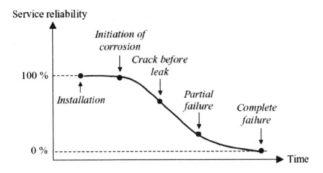

Figure 1.5 Deterioration process of underground pipes (Misiunas, 2006)

Deterioration leads to leakage. Higher leakage rates reduce water pressure and quantity at the supply point. Moreover, the infiltration of ground water can result in risk of water pollution, which increases the risk of likelihood of discoloured water and outbreaks of water borne disease. In the wastewater system, sewerage overflows create environmental problems and can potentially pollute receiving waters. Higher rate of water losses and lowered flow pressures significantly affect the designed performance level. Frequent failures of pipes increase the frequency of system maintenance and the need for emergency replacements.

Many sources of uncertainty exist in the deterioration of infrastructure systems. The deterioration and breakage processes of underground infrastructure systems are governed by static, dynamic, or multiple factors. The static factors include pipe characteristics, such as pipe material, diameter, wall thickness, and backfilled soil. The dynamic factors are age, temperature, soil moisture, soil electrical receptivity, bedding condition, dynamic loading, and operational factors (see Kleiner and Rajani, 2001). These multiple factors quite often exhibit a combined effect; for example, corrosion may have weakened a pipeline and excessive pressure (either internal or external) will then cause a pipe break (Seica and Packer, 2004). These factors cause water pipe deterioration mainly in two ways: i) structural deterioration diminishes the water pipes' ability to withstand various types of stress, and ii) deterioration of the inner surface of the pipes occurs due to internal corrosion. Many of those factors are uncertain making prediction difficult as many cases lack the required data for analysis.

iv)   *Socio-economic changes*

Changes in socioeconomic factors lead to an increase in water consumption and demand for better quality of service, and ultimately impacting the performance of UWS.

Many studies revealed a strong correlation between rate of water consumption and a number of socio-economic changes including economic development, lifestyle changes, higher levels of education, and lower water prices (Schneider and Whitlatch, 1991; Clarke et al., 1997; Bradley, 2004). A link has also been identified with types of housing and their sizes as well as acceptability and market penetration of water-efficient appliances. For example, water demand is increasing in many developing countries, particularly in China, Brazil, India, and South America, where economic growth is very high in recent years and is expected to continue to grow over the coming decades. Thus, increased socio-economic changes will have direct pressures on water consumption patterns and the water quality standards to be met.

Socio-economic changes such as willingness to pay and increased public participation in water companies have direct and indirect impact in UWM. Willingness to pay reflects consumers' trust in their service providers. It has been observed that people are willing to pay more for improved water services if there is transparency in the management decisions and investments of the utility provider. In contrast, consumer's willingness to pay reduces when the supplier is in the private sector (e.g. WTP studies in the UK, Argentina, and Sri Lanka). Consumer willingness is near zero when the supplier appears to be wasteful or profiting unfairly. Sometimes, despite the service structure and quality, affordability can limit a consumer's willingness to pay for water services (Raje et al., 2002).

A clear correlation exists between living standards and socioeconomic improvement to the quality of infrastructure services and increase in water demand. However, other factors, such as availability of water resources — if water resource is scarce or surplus; the pricing structure —if the price is block rated or fixed rated; the types of utility suppliers —if the utilities are private or public, will influence the water demand (Alcamo et al., 2007); which are uncertain.

1.3   Present State of Knowledge

From the above brief review, it is clear that UWS are highly vulnerable to uncertain future change pressures. The sources of uncertainties stem from prediction science to impact analysis. The decision making framework for long-term strategic planning requires a detailed understanding of likely and potential risks resulting from internal and external change future pressures (Hall and Solomatine, 2008; Wilby et al., 2009). A 'stationary approach' of infrastructure planning and decision-making where the future is assumed as the continuation of historical observations will not work (IPCC, 2012;

Hall et al., 2012; Beven and Alcock, 2012). Therefore, there is a need for a systematic framework to analyse the uncertainty and risks associated with UWS.

Planning for a long-term future by analysing the risks is not a new approach in UWM. However, most of the existing risk assessment frameworks do not consider uncertainty analysis. For example the risk assessment framework for human health (NRC, 1983), the ecological risk assessment framework the integrated risk assessment framework (WHO, 2001), and the cumulative risk assessment framework (EPA, 2003) to analyse risks to human health and the environment posed by chemical pollutants. Most of these frameworks are generic and applicable only for the health-based risk assessment (SuterII, 2007). In analysing a complex system like UWS, it is necessary to analyse the interactions between events and the associated uncertainties (Lindhe et al., 2009)

In order to address new environmental challenges, there has been gradual modification on the existing risk assessment frameworks. For example, a risk assessment framework for civil engineering facilities (Faber and Stewart, 2003), an environmental risk assessment framework that assesses the impacts of climate change (Jones et al., 2007), and a risk assessment framework for ecological risk management in the irrigation industry (Hart et al., 2005). Others also purpose new risk assessment frameworks to address multiple sources of hazards in a complex system. Linkov et al. (2006) proposes a new framework that incorporates multi-criteria decision analysis and adaptive management methods; Ashley et al. (2007) proposes an analytical framework to design an adaptable urban system that addresses climate change, urban form, and uncertainty. Ayyub et al. (2007) presents a quantitative risk assessment and management framework that supports strategic asset-level resource allocation decision making for critical infrastructure and key resource protection. However, none of these risk assessment frameworks adequately address uncertainty issues (see Chapter 2 for a critical review on risk assessment framework).

Various theories exist for uncertainty analysis, such as probabilistic analysis (Apostolakis, 1990); probability bound analysis combining probability and interval analysis (Ferson and Ginzburg, 1996); imprecise probability (Walley, 1991); random sets in two forms as proposed by Dempster and Shafer (1976); fuzzy set theory and fuzzy measures (Zadeh, 1965; Klir 1987); Shannon entropy (Shannon, 1948); and possibility theory (Dubois and Prade, 1988; Dubois, 2006) for example. However, there is no clear guidance for choosing a particular theory for analysis (Pappenberger et al., 2005). Selection of an uncertainty analysis method depends on many factors including types of information available (sources of uncertainty), complexities of a model (i.e., number of data to be used for the analysis) and purpose of the analysis (i.e., decision analysis or forecasting).

UWS have multiple types of data sources. Some of the data are deterministic type, such as length, diameter, thickness, and other physical dimensions of infrastructure systems. Other data are statistical and probabilistic types, such as historical temperature, rainfall and runoff, population growth, surface water discharge, water demand, rate of wastewater generated, rate of system failure, water quality variation, runoff rainfall relation, etc. Additionally, the data could be imprecise and qualitative, such as ground water level, ground water pollution, infiltration rate, roughness of pipes, public behaviour, water leakage and losses, willingness to pay for the services, etc. Moreover, some of the data could be unknown, such as limited data availability about the system's physical characteristics, magnitude, and intensity of future extreme rainfall, droughts, wishful attack (malicious) on infrastructure systems, etc. In order to handle such a wide-ranging and complex set of data, a single theory and method of uncertainty analysis will not be sufficient. Therefore, for the faithful representation of available information, we need to use more than one theory for uncertainty analysis.

Moreover, UWS are multi-disciplinary and they have multiple interactions among the natural water cycles, sub-systems, and human activities. The decision making process, therefore, needs to address the conflicting social, economic, and environmental objectives. Thus, the overall performance analyses of a system and risk management are multi-criteria decision analysis (MCDA) problems. It should be assessed against multiple conflicting objectives. The MCDA framework should be able to analyse the multiple types of data such as crisp, quantitative, and qualitative. There are several types of multi-criteria decision-making (MCDM) methods that have been applied in various fields (Romero and Rehman, 1987; Pohekar and Ramachandran, 2004; Hajkowicz and Higgins, 2008; Ananda and Herath, 2009). Still most of the multi-criteria analysis methods have limitations when dealing with data sets that are uncertain, ambiguous, and linguistic (Silvert, 2000; Roy, 2005). To address those shortcomings, there is need of analysing MCDA based decision-making framework for performance assessment of UWS and risk management in UWS.

1.4   Research Definition

This research has two main goals:(i) to develop a decision making framework, methodology, and tools for strategic planning of UWS, (ii) to demonstrate the developed methodologies in developed (data abundant) and developing countries (data limited) cases for the strategic planning of UWS.

The specific objectives of the research include:

1) To develop a risk assessment framework for urban water systems considering future change pressure and their associated uncertainties.
2) To develop a hybrid modelling method that allows uncertainties to be described and propagated through urban water models.
3) To develop a multi criteria decision making framework that is tailored for UWS operating under uncertainty.

4) To demonstrate the developed frameworks and methods by applying them to real-world UWS.

To achieve the goals; this research investigates the following research questions:

1) How should a risk assessment framework be structured to analyse risk and uncertainty for a strategic planning and decision-making process in UWM?

2) How can uncertainty be appropriately described, propagated, and modelled in UWS modelling and decision making?

3) How can one select a best strategy by analysing multi-objective and often conflicting criteria including multiple views of stakeholders?

4) How should the developed frameworks and methodologies be applied for decision-making in UWS?

The scopes of the research are: (i) it will not attempt to define, quantify, and apply sustainability in UWS; and (ii) the theories and discussions are focused to UWS but the application examples will not explicitly cover the UWS/urban water cycles.

1.5 Outlines of the Thesis

This thesis is organized into six Chapters as shown in Figure 1.6.

Chapter 2 provides a state of the art review on risk and uncertainty analysis and proposes a new risk assessment framework that can be applied for long-term, performance based decision-making in UWS planning. The Chapter introduces the basic concepts of risk, risk perception, risk aversion, risk assessment, and risk management prior to presenting a review on risk assessment frameworks used in UWS. The Chapter also discusses the different types and sources of uncertainty and the various methods for analyzing uncertainty in risk assessments. Finally, a risk assessment framework is proposed for analysing risk by considering the multiples types of uncertainty and multi-criteria based decision-making methods to analyse the performances of UWS and risk management.

Chapter 3 presents a hybrid approach to uncertainty analysis for risk assessment and UWS modelling. The Chapter discusses the different issues related to uncertainty description and propagation in UWS modelling. It then presents a review of the existing methods to hybrid approaches to uncertainty analysis and discusses their application potential in UWS. This research proposes a hybrid approach to uncertainty analysis where one form of uncertainty information is transformed into the other (i.e., probabilistic to fuzzy form and vice versa) and propagated homogenously. The homogenous uncertainty propagation is processed either by a probability theory based sampling technique or a fuzzy set theory based technique. The Chapter also presents the risk assessment methods based on fuzzy set theory considering the likelihood of risk of failure of performances and the consequences of failure. The proposed hybrid approach of risk assessment methodology and algorithms are operationalized for future water availability analysis in Birmingham, UK in the year 2035.

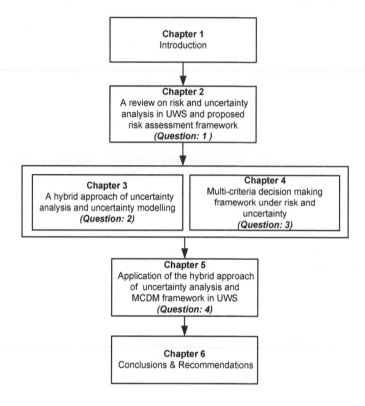

Figure 1.6 Outline of this thesis

Chapter 4 develops a multi-criteria based decision making framework for performance assessment and risk management in UWS under uncertainty. After an introduction to Multiple Criteria Decision Analysis (MCDA), commonly used MCDA methods that are applicable for performance assessment and risk management in UWS are reviewed. A MCDA framework and method based on fuzzy set theory is developed that meets the shortcomings of the existing methods of MCDA used in UWS. The developed framework and methods have been applied for the performance assessment and strategic planning of main urban infrastructure systems in Kathmandu, Nepal.

Chapter 5 demonstrates the hybrid approach of uncertainty analysis and multi-criteria decision-making framework in UWS in Mbale town, Uganda. The case study analyses the water availability for the next 20 years based on limited data cases. Both fuzzy membership functions and probability distribution functions are used to describe the uncertainty. The risk of water availability in Mbale is

14

analysed comparing the demand and supply situation in year 2032. The fuzzy set theory based multi-criteria analysis method developed from this study has been applied for the risk management. The results are also compared with the recent work of a World Bank study.

Chapter 6 presents the major conclusions and recommendations from this research.

Appendices: Probability equations, fuzzy set theory, fuzzy arithmetic, and risk modelling are presented at the end of the thesis.

Chapter 2   A Review on Risk and Uncertainty Analysis and Proposed Risk Assessment Framework

This Chapter presents a state of the art review on risk and uncertainty analysis and proposes a new risk assessment framework for a strategic decision-making in UWM. It firstly introduces the commonly used terminologies in risk assessment, such as risk concept, risk perception, risk aversion, risk assessment, and risk management. It then presents a brief literature review on existing risk assessment approaches that are applicable in UWS. The Chapter also defines the uncertainty and differentiates it from risk analysis. The types and sources of uncertainty analysis are presented and discussed. It further reviews different methods of uncertainty analysis and discusses the probability and fuzzy set theory-based methods of uncertainty analysis. Finally, it proposes a new risk assessment framework, discusses the developed methodologies, and draws some conclusions. This Chapter is based on the papers: Khatri and Vairavamoorthy (2009); Khatri and Vairavamoorthy (2011a); Khatri and Vairavamoorthy (2011b); Khatri and Vairavamoorthy (2013b).

2.1   Risk Assessment and Risk Management

2.1.1 The definitions and concept of risk

The word *risk* originates from the Greek word "*rhiza*", which refers to the hazards of sailing around a cliff (Covello and Mumpower, 1985). While the term has a long history of use, there is currently no commonly accepted definition of risk in either science or public understanding (Renn, 1998). Instead, risk is defined by particular sectors and uses. For example, in economics, risk applies to gains and losses from investments and is generally associated with a deviation from the expected return or defined as the quantifiable likelihood of loss. In the insurance sector, risk is treated as expected loss (Jonkman, 2007). In psychology, risk is defined as "the lack of perceived controllability", a "set of possible negative consequences", or "fear of loss" (Vlek, 1996). Further, the concept of risk adapted by social scientists considers risk as a socially constructed contextual notion that accounts probability of undesired consequences and the seriousness of possible undesired consequences.

In the natural sciences, risk is usually defined in terms of hazard and vulnerability. *Hazard* refers to the sources of possible damage or injury, or alternatively to something that causes risk. *Vulnerability* relates to potential consequences in the case of a hazardous event. Physical science and engineering also define risk interchangeably with words such as chance, likelihood, and probability to indicate that people are uncertain about the state of the activity, item, or issue under consideration. In this field, a widely used definition of risk is the combination of the probability or likelihood of an event and its consequences.

$$Risk\,(R) = Likelihood\,of\,Risk(L) \times Concequence\,(C) \qquad (2.1)$$

If number of events n with occurrence probabilities P_i and consequences C_i result from an activity, the total risk associated with the activity is the sum of the risk of the individual events.

$$R = \sum_{i=1}^{n} P_i C_i \qquad (2.2)$$

Generally, the notion of risk is comprised of two or three elements: the identification of failure or damage scenarios (what can go wrong?) as well as their likelihood and potential consequences. Thus, risk is a set of hazard events, their likelihoods, and their consequences (Kaplan and Gerrick, 1981), as in Equation 2.3.

$$R = <S_i, P_i, C_i> \qquad (2.3)$$

where, S_i = a scenario of events that lead to hazard exposure, P_i = the likelihood of scenario i, and C_i = the consequence (or evaluation measure) of the scenario i, e.g., a measure of the degree of damage or loss

Risk is both a descriptive and a normative concept (Renn, 1998). It relates to the analysis of cause-effect relationships that may be scientific or subjective in nature. Risk is also associated with events or actions that may be small or large—from household water pipe failure to damming in a sea. Total risk is the combined result of "how much or how often" and "some risk per unit of action or per event" Figure 2.1 shows the position of risk as a product of likelihood and consequences. A hazard with a higher probability but lower potential consequences to a system would have a similar risk compared with a hazard with a smaller magnitude of probability but larger potential consequences. It implies that the set of risk scenarios, product of likelihood, and consequences of failing events determines the risk magnitude.

This research defines risk as a function of the probabilities and consequences of a set of undesired events—a definition that has been used both in physical science and engineering. The particular risk of interest in this research is to develop a decision-making framework that allows the analysing of undesirable events in UWS, their likelihoods of occurrence, and their consequences due to the key future change pressures (described in Chapter 1). The term "undesirable events" signify a violation of the minimum expected levels of service in a system. Thus, a risk in a UWS is perceived if internal or external change pressures have violated the system's performances. System performances refer to the minimum designed level of services that humans or the environment depend on, potentially including water availability, water quality, or acceptable minimum failure rate of water systems.

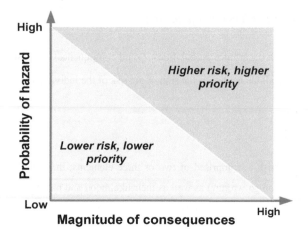

Figure 2.1 Risk definition as a function of likelihood of hazard and consequences

2.1.2 Risk perception and risk aversion

Risk perception is characterized as a subjective and personal representation of a risk. Thus, risks induced by a particular event are perceived in terms of their potential to impact people and the environment, particularly the extent of the impact—number of people affected or extent of an area impacted, and the duration of those impacts. Risk also depends on how soon or how likely an impact is expected to occur. For example, a flood event is perceived differently from one person to the next due to a variety of socio-economic and other local factors. People's perceptions are further influenced by their trust in experts and by their own ways of understanding, their own experiences, and their ability to imagine a risk's consequences (Slovic, 1987; Vlek, 1996).

Day Water (2003), however, describes risk perception entirely differently in terms of a few dichotomous factors such as: (i) Is this an individual risk or are more people undergoing the same risk? Is it a known risk? (ii) Do people have any past experiences with this particular risk and its consequences, or it is unknown to public, brought about by media? The worst fear is the fear of the unknown. (iii) Is this a voluntary risk taken on by an individual or group undergoing an activity that implies the risk? (iv) Is the risk involuntary, in that the individual or group would have avoided it had they had the choice?

Since UWS are public utilities, the interactions among the systems, environment, and people are diverse. Risk judgment and risk acceptability also depend on a number of other factors such as (see in reference in Crouch and Wilson, 1982):

- *Immediacy of effect*: Are the consequences immediate or delayed? Is there a latency period? If so, how does it affect perception of the risk situation?

- *Availability of alternatives*: Are alternatives available that provide similar benefits with reduced risks? If alternatives are available, perception and acceptance of risk may be easier.

- *Knowledge about risk*: How well can risk analysts determine the risk? How large are the uncertainties? How well do persons exposed to the risk understand the risk, its probability, consequences, and uncertainties?

- *Necessity of exposure*: How essential is exposure to the risky situation?

- *Familiarity of risk*: Old and familiar risks are more easily accepted than new or strange ones, particularly those that cannot be seen.

- *Chronic versus catastrophic risk*: Is the risk situation frequent or is it very rare with large numbers of fatalities possible?

- *Distribution of risk*: Is the risk situation widespread? Does it affect average persons or sensitive persons?

- *Potential for misuse or use*: If the technology is misused, do the risks increase significantly? How conceivable is misuse?

People exposed to risk may be individuals or a group of individuals sharing the risk together. Therefore, the behaviours of each individual are reflected in risk perception. Generally, personal temperament, personal values, social and cultural background, gender, decision-making ability, and education can influence the perception of a risk. However, most people perceive risk as a multidimensional phenomenon and integrate their beliefs with respect to the nature of the risk, the cause of the risk, the associated benefits, and the circumstances of risk-taking into one consistent belief system (Renn, 1998). Thus, those factors, which are more social, cultural, and psychological, are more likely to govern risk perception than quantified risk in socio-technical systems like UWS.

An individual's attitude toward risk is *risk aversion* and is related to risk perception. Most people are risk-averse; that is, they are willing to incur some cost to avoid risk. However, there is a wide range of degrees of risk aversion. People may be risk-averse if the potential losses are high, or risk-prone if the potential gains are high.

Risk aversion concerns the aversion to risks that cause large societal disruptions, such as accidents with multiple fatalities. Generally, one flooding event that causes 100 days of service disruption is perceived as more dreadful (and less acceptable) than 100 floods with one fatality each. This is the characterization of risk aversion. Risk aversion also refers to an aversion towards large consequences in general, including wide-scale economic loss. For example, the services disruption for 1,000 people in a single failure event would be valued as worse than service disruption of 1 person in 1,000 events.

From the viewpoint of UWS, risk aversion is the control action taken to avoid or eliminate a risk. It modifies activities to reduce the magnitude and/or frequency of adverse effects and reduce the vulnerability of exposed people and property. Risk aversion is also related to activities for development and implementation of adaptation and mitigation measures. Risk perception and risk aversion are thus major characteristics for risk analysis and management, especially for public utilities. However, this research focuses on the technical aspects of risk, and risk associated with social issues is beyond the scope of this work.

2.1.3 Risk assessment and risk management

There are not any clear definitions and distinctions between risk assessment and risk management that have been well accepted in risk literature. Haimes and Leach (1984) define risk assessment (RA) as the overall process of risk identification, quantification, evaluation, acceptance, aversion, and management. Renn (1998) explains RA as the scientific process of defining the components of risk in precise, usually quantitative terms. RA consists of calculating the probabilities for (un)wanted consequences, and aggregating both components by multiplying the probabilities by the magnitude of the effects.

Willows & Connell (2003) describe RA as the process of establishing information concerning hazards, and the exposure and vulnerabilities of defined receptors. Cowel et al. (2002) emphasizes the risk assessment as an estimate of the likelihood and severity of harm associated with a product, process, activity, agent (such as pollutants in different media), or event. Therefore, RA has been regarded as environmental management tools rather than concepts. Environmental Protection Agency (2004) defines RA as a process in which information is analysed to determine if an environmental hazard might cause harm to exposed persons and ecosystems. EPA uses RA as a tool to integrate exposure and health effects or ecological effects information into a characterization of the potential for health hazards in humans or other hazards to the environment.

One well refereed definition on risk assessment and risk management is given by Rowe (1977). Rowe defines risk assessment as the scientific process of quantifying risk, while risk management is the managerial response based on the resolution of the various policy issues, such as acceptable risk. Risk management decisions are made by considering risk assessment within the context of political, social, and economic realities. Risk assessment includes risk determination and risk evaluation, whereas risk management includes risk assessment and risk control (Figure 2.2). Risk determination involves the related process of risk identification and risk estimation. Risk identification is the process of observation and recognition of new risk parameters, or new relationships among existing risk parameters, or perception of a change in the magnitudes of existing risk parameters. Risk evaluation is a process of developing acceptable levels of risk to individuals, groups, or society as a whole. It involves the related process of risk acceptance and risk aversion. Risk acceptance implies that a society is willing to accept some risks to obtain a gain of benefit if the risk cannot possibly be avoided or controlled. The acceptance level is a reference level against which a risk is determined and the compared. If the determined risk level is below the acceptance level, the risk is deemed acceptable. If it is deemed unacceptable and avoidable, further steps may be taken to control it by risk management. Further discussion based on Figure 2.2 can be found in Rowe (1977).

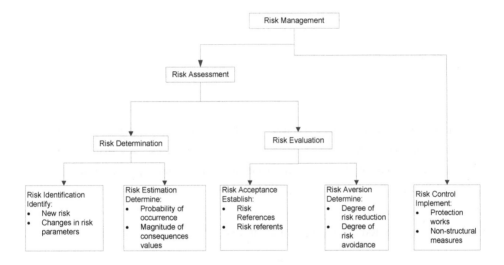

Figure 2.2 Steps in risk management (Rowe, 1977)

Risk assessment is a broad term that encompasses a variety of analytic techniques that are used in different situations depending upon the nature of the hazard, the available data, and the needs of decision makers. Therefore, risk assessment techniques can be quantitative, semi-quantitative, or qualitative. Both techniques can be used to describe the knowledge of risk where probabilities can be estimated with some level of confidence. Qualitative techniques are particularly useful in circumstances where knowledge of the probabilities is lacking. Risk assessment may therefore involve the combination of *qualitative* and *quantitative* information (Willows and Connell, 2003).

This research uses risk analysis and risk assessment as a synonymous term. As defined by Rowe (1977) and Willows & Connell (2003), risk assessment is considered the scientific process of quantifying risk, while risk management a response to reduce the identified risk in an acceptable level. Thus, risk assessment involves consideration of sources of potential impact on systems performances, assessment of likelihood, and the consequences of performance failures. Risk management evaluates which risks identified in the risk assessment process require management attention and selects and implements the plans or actions to ensure that those risks are controlled.

2.1.4 A review on risk assessment approaches

Existing research describes several approaches to address the various sources of risk in a complex system. Haimes et al. (1998) propose a risk assessment for reducing risks in water systems caused by wilful human attacks. They develop a hierarchical holographic model (HHM) based on 15 major categories (already developed in Haimes (1981). The HHM covers a range of components in water supply systems: physical, temporal, maintenance, organizational, management, resource allocation, supervisory control and data acquisition (SCADA), system configuration, hydrology, geography, external factors, system buffers, contaminants, and quality of surface and ground water. Each category is further divided into detailed components that are easier and more manageable to evaluate. This framework presents redundancy, robustness, and resiliency as important concepts for responding to risk (see, Haimes et al., 1998).

White et al. (1999) use a hierarchical model to reduce complexity in water distribution systems (Figure 2.3) and analyse the risk in a component level. In this model, a system is viewed as a collection of subsystems called 'regions' (collections of continuous water pressure zones). A specific region is then divided into 'areas' which are further reduced to smaller subsets called 'locals'. In this way, locals can be further divided into more detailed components that include pipes, pumps, valves, etc., down to the lowest levels of the hierarchical structure.

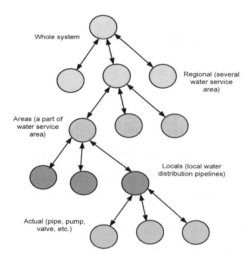

Figure 2.3 Hierarchical model for organising simulation assessment of a water supply system (White et al., 1999)

Ezell et al. (2000) propose a vulnerabilities-based risk assessment method that assesses vulnerability in terms of exposure and access control. Based on vulnerability analysis and expert opinion, a wilful water contamination attack scenario is developed and then modelled using an event tree. Vulnerability in a system is quantified through a detailed system-decomposition that uses the hierarchical method proposed by Haimes (1981). Vulnerabilities of a system are subjectively assessed by summing the product of accessibility and degree of exposure of components. Figure 2.4 presents the Ezell et al. (2000) method applied to water distribution systems.

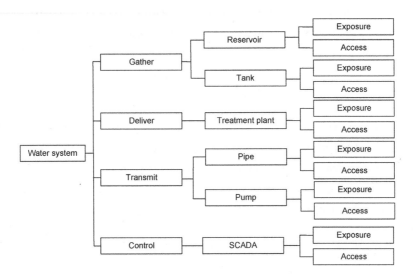

Figure 2.4 Hierarchical structure for vulnerability assessment of water supply system (Ezell et al., 2000)

Kaplan et al. (2001) extend the Haimes et al's (1998) Hierarchical Holographic Modelling method with the Theory of Scenario Structuring (TSS) for identifying a set of risk scenarios. They link scenarios with the "set of triplets" definition of risk (i.e., what can go wrong?; what is its frequency/probability?; what are the consequences?), in which the scenario set is a part of the definition. They emphasize that building a risk scenario is an important step in risk analysis. Thus, TSS is a comprehensive treatment of the process of finding, organising, and categorizing the set of risk scenarios.

Kleiner (1998) classifies water quality failures in distribution networks into five major categories: (i) intrusion of contaminants, (ii) re-growth of microorganisms, (iii) microbial/chemical breakthrough, by-products, and residual chemicals from water treatment plants, (iv) leaching of pollutants from system components into water, and v) permeation of organic compounds from the soil through system components into the water. These categories are combined to find the overall risk of water quality failures. Sadiq et al. (2004) draw on Kleiner's (1998) classification to propose a fuzzy set theory-based hierarchical model that evaluates aggregative risk of water quality failure in a distribution network. Each risk item is partitioned into its contributory factors, which are also risk items, and each of these is further partitioned into lower level contributory factors. The unit consisting of a risk factor is called the "parent", and its contributory factors are called "children". If a risk element has no children, it is called a "basic risk item" and is evaluated in terms of the failure event's likelihood and consequence. The likelihood and consequence are defined using fuzzy numbers to capture vagueness in the qualitative definitions. These fuzzy numbers are subjectively determined. Then, a multi-stage aggregation is conducted to obtain the risks of water quality. Sadiq et al. (2007) applies this aggregative risk analysis approach as well to calculate water quality failure in a distribution network due to decontaminant intrusion, leaching and corrosion, bio-film formation and microbial re-growth, permeation, and water treatment breakthrough (including disinfection by products formation).

Kleiner et al. (2004) propose a new approach to model the deterioration of buried pipes using a fuzzy rule-based non-homogeneous Markov Process. This deterioration model shows the possibility of failure at every point along the life of a pipe. The possibility of failure, expressed as a fuzzy number, is coupled with the failure consequence (also expressed as a fuzzy number) to obtain the failure risk as a function of the pipe's age. The use of fuzzy sets and fuzzy techniques incorporates the inherent imprecision and subjectivity of the data, and propagates these attributes throughout the model.

Kleiner et al. (2006) present a method of managing risks of failure in large, buried infrastructure assets. The method translates a fuzzy condition rating of the asset into a possibility of failure. This possibility is combined with the fuzzy failure consequences to obtain the fuzzy risk of failure throughout the life of the pipe. This life-risk curve can be used to make effective decisions on pipe renewal, such as when to schedule the next inspection or condition assessment, when to renew a deteriorated pipe, or what renewal alternatives should be selected.

Yan and Vairavamoorthy (2003) develop a hierarchical model for assessing the conditions and failure risk of pipes in water distributions systems. This model considers several indicators affecting water pipes established from the historical data available in other publications. In this hierarchical structure, two groups of water pipe deterioration indicators are selected: physical and environmental. The physical indicators include basic pipe properties such as age, diameter, and material. Environmental factors include surrounding condition, soil condition, and road loading as contributing factors. This method results in a fuzzy number that represents the condition of a pipe (see Figure 2.5).

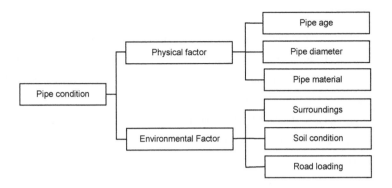

Figure 2.5 Hierarchical framework for pipe condition assessment (Yan and Vairavamoorthy, 2003)

Ezell (2007) presents an Infrastructure Vulnerability Assessment Model (I-VAM) to quantify the vulnerability of critical infrastructure in a medium-sized clean water system where vulnerability is a measure of the system's susceptibility to threat scenarios. In this process experts establish value functions and weights and provide an assessment of the protection measures of the system. This model accounts for uncertainty in measurement, aggregate expert assessment and yields a vulnerability density function. It has been claimed that I-VAM is useful to decision makers who prefer quantitative to qualitative treatment of vulnerability.

Recent risk approaches are focusing towards quantitative and comprehensive risk assessment techniques and consider risk parameters such as uncertainty, vulnerability, and resilience in their analysis. For example, Apel et al. (2009) present models for flood risk assessment that analyse hazards and vulnerability components in particular. The paper emphasizes that different methods and different levels of detail are required in flood risk assessment that vary according to assessment purposes. McIntyre and Wheater (2004) developed a tool for risk-based management of surface water quality called the Water Quality Risk Analysis Tool. The tool is based on simulation techniques like Monte Carlo Simulation that address uncertainties and has the capability to model organic pollution, phytoplankton, dissolved oxygen, various nutrients, toxic substances, floating and suspended oil, and total suspended solids. Similarly, Ayyub et al. (2007) present a quantitative technique of risk assessment for critical asset and portfolio risk from all sources of hazards.

Ashley et al. (2007) develop an analytical approach for local area drainage flood risk management that accounts climate change or urban form as a potential future change drivers, and identifies alternative adaptable solutions for local urban drainage systems. Wang and Blackmore (2008) investigate the ways that risk and resilience approaches may complement each other in the context of sustainable urban water system. Murray et al. (2008) argue that multiple methodologies for evaluating the vulnerability of critical infrastructure, particularly networks, is essential for deepening our understanding of the implications of unplanned events. Haimes (2009a) discusses the relationship between vulnerability and resiliency in risk analysis to make suggestions for improving system resiliency. Lindhe at al. (2009) develop an integrated and probabilistic risk analysis method for entire drinking water systems based on fault tree analysis. Their analysis includes situations of quantity failure (no water is delivered to the consumer) and quality failure (water is delivered but does not comply with water quality standards). They integrate hard data with subjective expert opinion to analyse failure probabilities. This analysis is simulated with the Monte Carlo technique for uncertainties propagation.

The review reveals a number of notable trends. Namely, four major types of risk assessment methods can be observed: (i) tree-based risk assessment, (ii) indicator-based assessment, (iii) step-based, and (iv) indicator and step-based methods. The commonly used tree-based techniques are fault tree analysis (FTA), even-tree analysis (ETA) and cause-consequences analysis (CCA). Indicator based assessment uses sets of risk indicators to measure the risk in a system. This method is therefore applicable for complex systems, which are often difficult to model. The step-based risk assessment relies on structured phases, or "steps", of risk assessment, such as system characterisation, hazard identification, likelihood or probability determination, etc. The indicator and step based methods use the combined approach of the two methods. Additional methods commonly used in risk assessment are outlined below in Table 2.1. Further discussion of methods are also available in Nicola & McCallister (2006); Ayyub et al. (2007); Aven (2010).

Table 2.1 Some examples of risk assessment methods and their scopes

Risk assessment method	Scope	Type of analysis
Safety/review audit	To identify equipment conditions or operating procedures that could lead to a casualty or result in property damage or environmental impacts	Qualitative type: (e.g., Cacciabue, 2005) for safety audit of a large railway organisation.
Checklists	To ensure that organizations are complying with standard practices. Commonly used in project management.	Qualitative type: (e.g., Ahmed et al., 2007) for risk management in projects.
"What-If" analysis	To identify hazards, hazardous situations, or specific accident events that could result in undesirable consequences.	Qualitative type: (e.g., Mythen and Walklate, 2008) for Terrorism, Risk and International Security.

Hazard and operability Study (HAZOP)	To identify system deviations and their causes that could lead to undesirable consequences, and to determine recommended actions to reduce the frequency and/or consequences of the deviations.	Qualitative type: (e.g., Duijne et al., 2008) quantified methods for risk assessment.
Probabilistic risk Analysis (PRA)	A methodology for quantitative risk assessment developed by the nuclear engineering community. This comprehensive process may use a combination of risk assessment methods.	Quantitative type: (e.g., Ezell et al., 2010) for Terrorism Risk.
Preliminary hazard analysis (PrHA)	To identify and prioritize hazards leading to undesirable consequences early in the life of a system. It determines recommended actions to reduce the frequency and/or consequences of prioritized hazards. This is categorized as an inductive[1] modelling approach.	Qualitative type: (e.g., Huang et al., 2008) for a nuclear power plant.
Failure modes effects analysis (FMEA)	To identify the components affected by failures and the impacts on the surrounding components and the system. This has been categorized as an inductive modelling approach.	Quantitative type: (e.g., Liu et al., 2012) a review on FMEA.
Fault tree analysis (FTA)	To identify combinations of equipment failures and human errors that can result in an accident. This has been categories as a deductive modelling approach.	Quantitative type: (e.g., Lindhe et al., 2009) for fault tree analysis for probabilistic risk analysis of drinking water systems.
Event tree analysis (ETA)	To identify various sequences of events both failures and successes—that can lead to an accident. This has been categorized as an inductive modelling approach.	Quantitative type: (e.g., Bucci et al., 2008)

The research being developed in this field is clearly substantial. However, there are significant limitations to the developed methods.

- Most of the developed approaches focus on only one aspect of risk. For example, Haime's (1998) method only assesses threats of human-related wilful attacks and does not address uncertainty in risk analysis; Yan's (2003) method assesses only actual pipe conditions and.

---

[1] Inductive methods analyses different end states of a systems due to some event. The event in risk analysis is often fault or failure. This approach includes PrHA, FMEA and ETA. Further detailed discussions are available in the cited sources.

does not include important probabilistic information in indicator; Sadiq's (2004) method is specifically developed to assess water quality problems in water distribution networks and cannot be applied to other sources of hazards; and Kleiner's (2006) method is designed for risk management of large buried infrastructure assets only

- Most of the developed approaches are unable to account for the interdependencies of a lifeline system and to calculate the cumulative risks in such integrated systems. In a lifeline system, the system will fail if any one of its components fails; water supply, for example, can fail as a result of failures in transmission pipes, treatment works, or distribution systems. While Lindhe's (2009) method presents a framework for water supply systems, the framework only applies to failures in the water supply system itself. This technique is also not applicable if there are imprecise type of data, which occurs in most real cases.

- Some approaches that consider system complexity do not consider uncertainty. For example, both Haime's (1998) method, White's (1999) method, and Ayyub's (2007) method deal with complexities of a system, but do not explicitly consider uncertain information associated with the basic parameters in their frameworks. As described earlier, uncertainty is an inherent component of complex systems that must be considered.

- Some approaches employ probability theory while others use fuzzy set theory to capture their data sources. For example, McIntyre (2004), Ashley (2007), and Lindhe (2009) use probability theory based approaches, whereas Kleiner (1998), Yan (2003), and Sadiq (2007) use fuzzy set theory. These theories are only appropriate for homogenous types of data and will not be applicable for combing qualitative and quantitative data sets. To combine qualitative and quantitative data, a hybrid approach or semi-quantitative analysis is necessary.

- Some approaches describe issues of vulnerability (Haimes, 2009), resiliency (Wang and Blackmore, 2008), and future change pressures (Ashley et al., 2007), but do not provide a clear methodology for how these issues can be operationalized to address the multiple sources of future change pressures affecting UWS.

Though this review is by no means exhaustive, it clearly identifies an existing gap in current risk assessment methods. Current approaches are sufficiently unable to consider multiple sources of hazard due to future change pressures and are not equipped to capture and address uncertainties inherent in complex systems. As such, existing risk assessment frameworks are insufficient for effectively analyzing complex, integrated, and dynamic systems like UWS. Uncertainty analysis is one of the most important steps in the risk analysis. This research develops a new risk assessment framework addressing those shortcomings, particularly incorporating the uncertainty analysis in risk assessment and multi-criteria based risk management for the decision-making. The risk assessment framework and methodologies is presented in the section following the risk and uncertainty analysis, which presents methods of uncertainty analysis for risk assessment in UWS.

2.2 Risk and Uncertainty Analysis

2.2.1 Risk and uncertainty

Uncertainty is often defined with respect to certainty. *Certainty* implies that the analyst has appropriate information to describe, prescribe or predict deterministically and numerically a system, its behaviour or other phenomena (Zimmermann, 1987). Uncertainty generally results from insufficient information, which can be reduced if more information is available. For example, future flood hazards posed by a changing climate involves uncertainty in their magnitude and intensity which cannot be presented deterministically or accurately (Willows and Connell, 2003).

In literature, uncertainty has been distinctly differentiated from risk. Knight (1921) establishes this distinction in his seminal work, "Risk, Uncertainty, and Profit":

> ... *Uncertainty must be taken in a sense radically distinct from the familiar notion of Risk, from which it has never been properly separated.* ... The essential fact is that "risk" means in some cases a quantity susceptible of measurement, while at other times it is something distinctly not of this character; and there are far-reaching and crucial differences in the bearings of the phenomena depending on which of the two is really present and operating. ... It will appear that a measurable uncertainty, or "risk" proper, as we shall use the term, is so far different from an un-measurable one that it is not in effect an uncertainty at all.

Knight describes risk as imperfect knowledge in which possible outcome probabilities are known, that is, in which the uncertainty is measurable, whereas uncertainty exists when these probabilities are unknown, unable to be measured.

Other key terms such as variability, error, accuracy, precision, ignorance, and confusion are different than uncertainty, though they are often used synonymously. *Variability* is a distinct property inherent to any system whose makeup consistently differs in some significant way, as in daily rainfall or the concentration of chemicals in an effluent. Variability, unlike uncertainty, can be observed, estimated, and directly verified, but cannot be reduced (SuterII, 2007). For example, if we observe stream flow over a particular spatial or temporal sampling frame, we can estimate the distribution of variable parameters of the flow, and we can verify our estimates by taking measurements. However, we cannot predict the minimum flow in the stream because the system's meteorology and hydrology contain variability and because of uncertainties concerning the applicability of present available data to the future. The joint influences of variability and uncertainty contribute to future unpredictability. This combination of variability and uncertainty is often termed *total uncertainty*.

Uncertainty thus relates to the availability of information and the ability to make predictions. Certainty implies accurate prediction, while uncertainty is a measure of the inability to accurately predict or define. A remark made by former Defense Secretary Donald H. Rumsfeld's "Known and Unknown" in 2002 about the Iraq war signifies the relation between information in hand and uncertainty: Reports that say that something hasn't happened are always interesting to me, because, as we know, there are known knowns; there are things we know we know. We also know there are known unknowns; that is to say we know there are some things we do not know. But there are also unknown unknowns — the ones we don't know we don't know.

Risk assessment in an UWS inevitably involves a process with many known unknowns as well as many unknown unknowns. This stems from the natural variability of systems, the complexity of social interactions in built systems, and the technological complexities in system models and model processing. In this respect, the conditions that must be met for certainty in risk assessment in the context of climate change are given in Figure 2.6. As shown, a good understanding of risk will be possible in cases where the climate is unchanging, where there is availability of good historical data for analysis, and where good impact modelling and prediction for the short term are available. On the other hand, a poor understanding of risk would be encountered, for example, when there is rapid climate change, lack of sufficient data and when long-term forecasts are desired, as shown in the bottom left corner of the Figure 2.6.

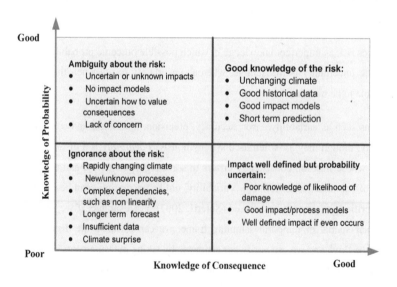

Figure 2.6 Environment for risk definition and quantification (Willows & Connell, 2003)

Uncertainty is a manifestation of information deficiency and describes the quality of knowledge concerning risk that needs to be faithfully represented, propagated and presented for the decision-making process. Thus, uncertainty is a state of knowledge between certainty, which is perfectly known, and the present state of information, which may be partially known or completely unknown.

2.2.2 Types of uncertainty

The literature reflects a number of classifications for types of uncertainty. Yen and Ang (1971) define uncertainties as *objective* or *subjective*, where *objective* uncertainty arises from random processes and are deductible from statistical samples, while *subjective* uncertainty is caused by imprecision, where no quantitative information is available. Der Kiureghian (1989) presents types of uncertainty as *reducible* or *irreducible*, in which uncertainties associated with unknowable things are *irreducible*, and uncertainties associated with knowable things that are currently unknown are *reducible*.

Uncertainties are also classified according to *vagueness, ambiguity,* and *underspecificity* (Ayyub and Chao, 1998; Klir and Folger, 1998; Colyvan, 2008). *Vagueness* uncertainties are derived from vague statements like "water supply is irregular". In this statement, the meaning of "irregular" is unclear—its meaning may be context dependent as it may be supply interruption for several days or a few hours in a day. *Ambiguity* uncertainty refers to the possibility of multiple outcomes for a process or system. *Ambiguity* may arise from the fact that a word can be used in more than one way in a given context. For example, the word "good" is ambiguous between different levels of services. This uncertainty is a non-crispness of belonging and non-belonging of elements in a set or in a notion of interest (Ayyub and Chao, 1998). Similarly, *underspecificity* is due to more generality. The statement that "there will be rainy days ahead" is an example of *underspecificity,* as it is uncertain how many days will be rainy, which day will be rainy, what time of the day will be rainy, or what will be the intensity and duration of rainfall.

Other uncertainty types include *randomness* or *fuzzy* types (classified based on their information sources) (Kaufmann and Gupta, 1991), as well as *aleatory* and *epistemic* uncertainty (Parry, 1996; Oberkampf et al., 2002; Hall, 2003; Helton et al., 2010). *Aleatory* uncertainty arises from a random process and is also referred to as *inherent* variability (Hall, 2003). *Aleatory* uncertainty is best-modelled using probabilities. *Epistemic* uncertainty results from lack of knowledge about an object under investigation. In this case, lack of knowledge refers to the capacity of the analyst to understand, measure, and describe the system under study. Furthermore, uncertainty may be due to ignorance where current knowledge is insufficient in describing the uncertainty. For simplicity, this research considers uncertainties in two classes as aleatory and epistemic. Further discussion on the types of uncertainty is available in Van der Klis (2003); Maskey (2004); Shrestha (2009).

2.2.3 Sources of uncertainty

Sources of uncertainties refer to location and causes of uncertainty. The major sources of uncertainty identified in the literature (Paté-Cornell, 1996; Krzysztofowciz, 2001; McIntyre et al., 2003; Maskey, 2004; Mannina et al., 2006; Beven and Alcock, 2011) are also applicable in the context of UWS. For example, uncertainty in climate change arises from two quite different sources: *incomplete* knowledge and *unknowable* knowledge. In the case of *incomplete* knowledge, future advances in climate science and computing technology will add to the knowledge, or in other words, future advancements will make the knowledge more complete as more information becomes known. *Unknowable* knowledge arises from the inherent indeterminacy of future human societies and the climate system. For example, societal actions are not predictable in any deterministic sense and always have to estimate future greenhouse gas emissions trajectories because of indeterminate scenario analysis (Nakicenovic et al., 1998).

Figure 2.7 presents the two sources of uncertainty: *variability* and *limited knowledge* that exists in a complex system (Van Asselt and Rotmans, 2002) In *variability*, the system or process under consideration has a random character and can behave in different ways, which causes uncertainty. This is also referred to as *objective* uncertainty, *stochastic* uncertainty, *primary* uncertainty, *external* uncertainty, or *random* uncertainty. It is noted that the types of uncertainties that appear on the right side of Figure 2.7 are sourced from the left side (represented by arrows). The middle block shows the level of indeterminacy staring from inexactness to irreducible ignorance generated from variability and their cause for the limited knowledge.

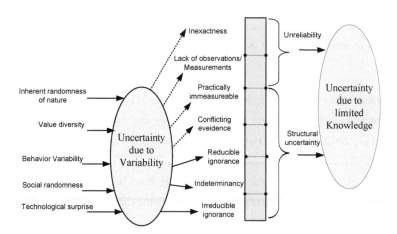

Figure 2.7 Typology of sources of uncertainty (Van Asselt and Rotmans, 2002)

Uncertainty due to *limited knowledge* is a property of the analyst performing the study and their general state of knowledge. This source of uncertainty is often categorized similarly to subjective uncertainty, incompleteness of the information, secondary uncertainty, or internal uncertainty. Limited knowledge partly results from variability and can range from *inexactness* to *ignorance*.

Uncertainty caused by *inexactness* results from lack of precision, inaccuracy, or measurement errors. Inexactness can be translated as "*we roughly know*" and includes a number of conditions affecting the roughly known. Lack of observation/measurements, for example, is a lack of data that could have been collected but were not. This is a situation of "*we could have known*". Practically *immeasurable* describes the condition in which missing data can be measured principally but not in practice, a situation of "*we know what we do not know*". *Conflicting evidence* describes a condition in which different data sets and observations are available but have competing interpretations. In this situation, "*we don't know what we know*". Similarly, *reducible ignorance* is a situation in which the analyst are unable to observe or theoretically possibly know, but may be able to in the future. Here, "*we don't know what we do not know*". *Indeterminacy* results when we understand a set of principles or laws but will never be able to fully predict or determine them. In this situation, "*we will never know*". Finally, *irreducible ignorance* describes a condition of uncertainty where the processes and interactions between processes under study cannot be determined by human capacities and capabilities. Here, the situation of "*we cannot know*" is applicable (see reference in-Van Asselt and Rotmans, 2002).

Sources of uncertainty can further range from unreliability to more fundamental, radical, structural, or systematic uncertainty. Uncertainty due to *unreliability* can be measurable or calculated, assuming that it stems from well-understood systems or processes. For this reason, it can be described quantitatively. On the other hand, *structural uncertainties* can only be roughly estimated. *Structural uncertainties* arise from conflicting evidence, ignorance, indeterminacy, and uncertainty due to variability.

For the purpose of risk assessment, identifying a source of uncertainty will signify where uncertainty needs to be treated (Abebe, 2004; Maskey, 2004). For complex systems such as UWS, there are four possible locations of uncertainty: uncertainty in the input data, uncertainty in the model parameters, uncertainty in the model, and uncertainty in the natural or operational process under study.

2.3   Approaches of Uncertainty Analysis

Uncertainty analysis is a process of the *description of uncertainty, propagation of uncertainty* and *analysis of output results,* as shown in Figure 2.8. In risk assessment, uncertainties in input parameters are described by a distribution function or interval, such as a probabilistic distribution function or membership function. The uncertain information is then propagated through a model by numerous analytical or random sampling techniques according to the information contained by the input parameters. Thus, identifying sources and types of uncertainty are important steps in risk analysis.

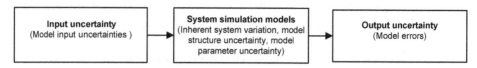

Figure 2.8 General process of uncertainty analysis

Different types of uncertainties require different methods of analysis. Identifying the ways that uncertainties manifest and then classifying the uncertainties into corresponding types according to their sources simplifies their analysis and modelling. The commonly used theories for analysing uncertainties include probabilistic analysis (Apostolakis, 1990); probability bound analysis combining probability and interval analysis (Ferson and Ginzburg, 1996); imprecise probability (Walley, 1991); random sets in two forms as proposed by Dempster and Shafer (Shafer, 1976); fuzzy set theory and fuzzy measures (Zadeh, 1965; Klir, 1987); Shannon entropy (Shannon, 1948) and possibility theory (Dubois and Prade, 1988; Dubois, 2006). Most of these theories are rich in content and detailed discussion on each of the theories is beyond the scope of this research. Further detailed discussion on different theories are available elsewhere (see, Klir and Yuan, 1995; Ayyub and Chao, 1998; Klir and Smith, 2001; Baudrit and Dubois, 2005; Dubois, 2006; Yao et al., 2011).

Both the probability theory and fuzzy set theory are the commonly used uncertainty analysis theories for risk assessment. These theories are equally applicable to capture and describe uncertainties associated with UWS. It is noted that this research applies both the probabilistic and fuzzy set theory approaches for uncertainty and risk analysis, both of which are further discussed in the following sections.

2.3.1 Probability theory

Probability theory represents uncertainty as random variables or a stochastic process. It is the best understood and most widely applied theory in handling uncertainty in risk assessment (Hall, 2003). A few examples of the probability theory based uncertainty analysis include structural failure risks (Haldar and Mahadevan, 2000); risk of natural disasters including floods (Apel et al., 2009); climate change impact studies to flood risk (Prudhomme et al., 2010); Earthquake risk (Ang and Tang, 1984); human health risks from exposure to contaminants (Khadam and Kaluarachchi, 2003); risk assessment of contaminated land (Öberg and Bergbäck, 2005); risk assessment of water supply system (Lindhe et al., 2009); least-cost design of water distribution networks (Babayan et al., 2005) and risk of water quality (McIntyre and Wheater, 2004) for a few examples. For further detailed discussion reader are suggested to view elsewhere (Paté-Cornell, 1996; Helton et al., 2003; Avena and Zio, 2010).

In probabilistic approaches of uncertainty analysis, probability density functions (PDFs) or probability mass functions (PMFs) of input parameters are developed from available data and propagated through an analysis model. The probability theory based propagation techniques are broadly classified into three categories: i) analytical method, ii) approximation method, and iii) sampling method.

i)   *Analytical method*

A common analytical method is the *derived distribution method,* also called transformation of variables method. It determines the distribution of random variable Y, which is related to another random variable, X whose PDFs or cumulative distribution functions (CDFs) are known, through the functional relationship $Y = f(X)$. The CDF of Y can be obtained as:

$$F_Y = F_X[f^{-1}(y)] \tag{2.4}$$

where $f^{-1}(y)$ represents the inverse function of f.

Then the PDF of Y, $F_Y(y)$ can be obtained by taking the derivative of $F_Y(y)$ with respect to y as:

$$f_Y(y) = \frac{dF_Y(y)}{dy} = f_X(f^{-1}(y)) \frac{df^{-1}(y)}{dy} \tag{2.5}$$

Equation 2.5 forms the basic building block of uncertainty analysis. Other methods and modelling techniques are based around this relation. However, this approach is applicable only for uncertainty having a single variable or linear type equation.

ii)   *Approximation method*

First Order Second Moment (FOSM) approximates uncertainty in terms of both the variance of FOSM approximates uncertainty in terms of the variance of system outputs from given values of mean and the variance of the uncertain input variables. It uses the first-order terms of the Taylor series expansion about the mean value of each input variable. This method is also referred to as *First Order Reliability Method, First Order Variance Estimation Method,* and *First Order Analysis.*

Considering the model output Y as a function of input random variable X as: $Y = f(X)$ the mean μ_Y and the variance σ^2_Y of Y are given by:

$$\mu_Y = \int_{-\infty}^{\infty} f(x) f_X(x) dx \tag{2.6}$$

$$\sigma^2_Y = \int_{-\infty}^{\infty} [f(x) - \mu_Y]^2 f_X(x) dx \tag{2.7}$$

In order to compute the mean (first moment) and variance (second moment) of Y using the above equation, information on the PDF $f_X(x)$ is required. In many cases, however, the probability density function $f_X(x)$ may not be known, and information about the mean and variance of X may be limited. Furthermore, even when $f_X(x)$ is known, the computation of the integrals in Equations (2.6 and 2.7) can be difficult and time consuming. Therefore, approximation methods like the FOSM would be practically useful to approximate the mean and variance of the output (Ang and Tang, 1975).

Moreover, FOSM is based on a number of assumptions: (i) that the relation between model output and input is linear or close to linear within the range of variation considered; (ii) that uncertainty about the model input is small (Coefficient of variance <02-0.25); and iii) that the first two moments of the probability distributions of the uncertainties exist (Van der Klis, 2003). Although this method is widely used in many civil engineering problems, it is difficult to apply in a complex modelling, which has problems with nonlinear functions of multiple variables. For further discussion of the methods, see (Zhao and Ono, 2001) and limitations and improvements of FOSM, see Maskey (2004).

iii) Sampling techniques of uncertainty propagation

Sampling methods provide the estimation of the probability distribution of an output, while approximation methods provide only the moments of the distributions (Haldar and Mahadevan, 2000). Monte Carlo and Latin hypercube sampling techniques are widely applied sampling methods in uncertainty analyse. Considering their application potential for UWS, Monte Carlo and Latin hypercube sampling techniques are further discussed in the following section. More detailed discussion on sampling methods are available in publication cited elsewhere (Helton et al., 2003; Helton et al., 2006).

a) Monte Carlo Simulation (MCS)

MCS is the most commonly used sampling method of uncertainty analysis. It relies on the central limit theorem to provide an overall assessment of the uncertainty in a prediction (Macdonald and Strachan, 2001) related to all uncertainties in the input parameters, regardless of interactions and quantity of the parameters. An MCS first assigns a PDF to each parameter under consideration. Values for all parameters are then randomly selected from their PDFs, and a simulation is undertaken repeatedly. After a number of simulations, the simulation results will form a distribution with an uncertainty. A few examples of MCS application in UWS include flood risk assessment (Apel et al., 2009); water quality risk analysis McIntyre and Wheater (2004); local area drainage flood risk (Ashley et al., 2007); risk analysis for drinking water system by fault tree analysis by Lindhe at al. (2009).

The MCS method can maintain the non-linear character of the model. It is applicable to process any types of the input uncertainties as long as they can be described in statistical terms. The output of the model is generated in a probability distribution functions that allows for complete statistical analysis of the results. Despite of a proven robust method of uncertainty analysis, it has some major disadvantages. For example, MCS requires a large number of simulations, and there is likely to be missing sampling in some areas. Additionally, it can be difficult to identify the distributions for many input parameters, and MCS is not applicable for uncorrelated random variables (Van der Klis, 2003). Further discussion on method of MCS modelling is presented in Chapter 3.

b) *Stratified Sampling and Latin Hypercube Sampling (LHS)*

In stratified sampling, the probability distribution of the input variable is typically divided into several strata of equal probability; one value is then chosen at random within each stratum. This method is an improvement over simple random sampling ensuring that samples from the whole distribution area are included.

Latin hypercube sampling method is an evolution of stratified sampling. In LHS, the region between 0 and 1 is uniformly divided into N non-overlapping intervals for each random variable. The N non-overlapping intervals are selected to be of the same probability of occurrence. Then, different values in the N non-overlapping intervals are randomly selected for each random variable, i.e., one value per interval is generated. This can be accomplished by initially generating N random numbers within the range [0,1]. The values are linearly transformed to the random numbers in the non-overlapping intervals for each random variable using the following equation (Susan et al., 1988; McKay, 1992; Ding et al., 1998):

$$u_i = \frac{u}{N} + \frac{(i-1)}{N} \tag{2.8}$$

where $i = 1,2,..., n$; u = a random number in the range [0,1]; and u_i = a random number in the ith interval

From the above equation, it is obvious that there is only one generated value that is randomly selected within each of the N intervals for each random variable. This is due to the following relationship:

$$\frac{(i-1)}{N} < u_i < \frac{i}{N} \tag{2.9}$$

where $(i-1)/N$ and $1/N$ are the lower and upper bounds for the ith interval.

Mckay et al. (1992) has shown that the stratified sampling procedure of LHS converges more quickly than the random sampling employed in MCS and other stratified sampling procedures. LHS results are more precise than MCS results based on the same numbers of model runs in linear and monotonic problems. However, in non-monotonic problems, LHS may or may not be more efficient than MCS (McKay, 1992; Ferson et al., 2004). LHS is a widely used techniques in water distribution system optimization problems (Kapelan et al., 2005) and water demand analysis (Khatri and Vairavamoorthy, 2009) where the numbers of input parameters are more. Some recent analysis concerning the LHS limitation are available in (Helton et al., 2003; Deutsch and Deutsch, 2012). Further detail on LHS modelling is presented in Chapter 3

Shortcoming of probabilistic approach

Although probability theory has been successfully used in many applications, it has some limitations. The major limitations are outlined below:

- Probability based methods need sufficient data to generate probability density functions or probability mass functions, which may not be possible in many field conditions. Additionally, if data is imprecise or insufficient (e.g., too few observations), the use of probabilistic models may be less reliable (Dubois and Guyonnet, 2011). Thus, the results will be sensitive to the amount of available information and simplifying assumptions (Ferson and Ginzburg, 1996).

- In many risk assessment modelling, some uncertainties are epistemic type. In such cases, it may not be possible to analyse the data using the probability theory. For example, roughness of underground pipes, quality of groundwater, contamination in receiving water, level of infrastructure services will be difficult to describe by PDFs. Epistemic uncertainty is represented by a standard distribution (e.g., normal distribution; uniform distribution) with a larger variance. A conclusion drawn from simplified distribution may be misleading (Dubois, 2010).

- Simple techniques to probabilistic uncertainty analysis, such as approximation method or First Order Second Moment method are not easily applicable to the analysis of complex systems. Simulation techniques are an alternative, but they require specialized skills and computing systems that are generally difficult to utilize in complex modelling and analysis.

- Both the First Order Second Moment and Second Order Reliability Method result in significant differences in their probability of failure predictions in asymptotic analysis of reliability, as shown by Adhikari (2005) and Maskey (2004). The difficulty in computing exceedance probability increases rapidly with the number of variables or dimensions.

2.3.2 Fuzzy set theory

Fuzzy set theory and *fuzzy measures* provide a non-probabilistic approach for modelling uncertainties associated with vagueness, imprecision, and inexactness due to human judgement (Ralescu and Ralescu, 1984; Zadeh, 2008). This approach has a broad range of application in risk-based decision-making, such as aggregative risk of water quality failure in a distribution network (Kleiner, 1998; Sadiq et al., 2004); the deterioration of buried pipes using a fuzzy rule-based non-homogeneous Markov Process (Kleiner et al., 2004) a hierarchical model for assessing the conditions and failure risk of pipes in water distributions systems (Yan and Vairavamoorthy, 2003); structural failure analysis (Klir and Yuan, 1995); geophysical, biological, and engineering systems analysis (Bardossy and Duckstein, 1995); and computing (Zimmermann, 2001). Further detailed review on analysis and design methods of model based fuzzy control systems is available in Feng (2006).

In classical fuzzy set theory, any element x of the universal set X can be classified as being either an element of some sub-set $A (x \in A)$ or an element of its complement, $x \in \bar{A}$, $(x \in \bar{A})$ that is, $(x \notin \bar{A})$. In other words, the transition for an element in the universe between membership and non-membership in a given set is abrupt and well defined, or *crisp*. Therefore, the membership either certainly belongs to the set (which is 1), or certainly does not belong to the set (which is 0). However, in many practical situations, the boundaries of sets are vaguely defined, and an element of a set can have a degree of membership other than one, where the transition between the membership and non-membership can be gradual rather than crisp. Zadeh (1965) first implemented this concept with the introduction of *fuzzy set theory*. A gradual transition of memberships is enabled by the "fuzzy" nature of fuzzy set boundaries, which are defined imprecisely and vaguely. This property of a fuzzy set makes the fuzzy set theory viable for representing uncertainty in a non-probabilistic form (Maskey, 2004).

Fuzzy set theory based uncertainty analysis method can handle uncertainties associated with vagueness, imprecision, or lack of information (Zadeh, 1983; Klir, 1987). The most widely used fuzzy set theory based analysis methods for handling uncertainties are the fuzzy extension principle and fuzzy alpha-cut technique. Some basic mathematical operations in fuzzy set theory are presented in Appendix II.

Fuzzy extension principle and fuzzy alpha-cut technique

The *fuzzy extension principle,* first developed by Zadeh (1975) and later extended by Yager (1986) enables the computation of the fuzziness of output-based fuzzy input variables, fuzzy mapping functions, or both. The *fuzzy extension principle* provides a mechanism for the mapping of the uncertain input variables defined by their membership functions to the resulting uncertain output variables (Maskey, 2004). In any function $Y = f(X)$, if the input X is deterministic or random, then output Y is also deterministic or random, respectively. In the same way, when X or its function is fuzzy, then Y is also fuzzy.

Consider any function f having several uncertain input variables $X_1, X_2, ..., X_n$ as

$$Y = f(X_1, X_2, ..., X_n) \tag{2.10}$$

If fuzzy sets $\tilde{A}_1, \tilde{A}_2, ..., \tilde{A}_n$ are defined by, $X_1, .X_2, ..., X_n$ such that $x_1 = X_1, ..., x_n = X_n$. Here, the output of such mapping is also a fuzzy set $\tilde{B}$. This is also called the image of fuzzy set $\tilde{A}$ on $X_1, X_2, ..., X_n$ under the mapping f, and it is found through the same mapping as

$$\tilde{B} = f(\tilde{A}_1, \tilde{A}_2, ..., \tilde{A}_n) \qquad . \tag{2.11}$$

where the membership function of the image $\tilde{B}$ is given by

$$\mu_{\tilde{B}}(y) = \max \left\{ \min \left[\mu_{\tilde{A}_1}(x_1), \mu_{\tilde{A}_2}(x_2), ..., \mu_{\tilde{A}_n}(x_n) \right] \right\} \tag{2.12}$$

The direct application of the fuzzy extension principle, especially for complex problems with many input variables, involves a computationally intensive procedure. Therefore, the *Fuzzy alpha-cut technique* is alternatively used (Abebe et al., 2000). The *alpha-cut of a fuzzy set* $\tilde{A}$ denoted as A_α is the set of elements x of a universe of discourse X for which the membership function of $\tilde{A}$ is greater than or equal to alpha; i.e.,

$$A_\alpha = \left\{ x \in X, \mu_A(x) \geq \alpha, \alpha \in [0,1] \right\} \tag{2.13}$$

The uncertainty represented by the fuzzy numbers can be propagated using alpha cut, which offers a convenient way of resolving fuzzy set into crisp set. Further discussion and modelling based on fuzzy alpha-cut has been proposed in Chapter 3.

Despite the wide applicability of fuzzy set theory in natural science and engineering, the following limitations and criticisms have been noted:

- Fuzzy set theory is commonly used for data sets where the bounds are not precise in contrast to interval analysis. It can be expected that the evaluation of these sets impart subjective uncertainties. Thus, fuzzy predictions are likely to deviate from exact values (Dubois and Prade, 1993).

- Fuzzy set theory is based on imprecise information and ordinal structure rather than additive structure (Dubois et al., 1996). Thus, the results of computations using these quantities may not be mathematically precise.

- In most cases, fuzzy set theory deals with qualitative data representing expert views and knowledge. For example, fuzzy evaluation for risk management requires that risk attributes, such as climate risk versus non-climate risk, or microbiological risk versus chemical risk, be assigned values of relative importance, which are determined by experts. However, experts may have differing views of risks due to bias. Choosing more experts for evaluation may reduce bias, but this remains a source of uncertainty. This procedure is likely to generate an interval rather than a single value. It needs complex mathematical approaches, such as gradient eigenvector and/or max–min paired elimination method, to solve such problems (Aven, 2010).

2.3.3 Possibility theory

Possibility theory primarily measures uncertainty of imprecision and handling of incomplete information, which arises in natural language and in measurements of physical systems (Dubois and Prade, 1998; Dubois, 2010). Recently it has been applied for dealing with the epistemic types uncertainties where uncertainties are described inn interval. Possibility theory emphasizes the quantification of the semantic, context-dependent nature of symbols – that is, *meaning* rather than *measure* of information. The guiding principles of the possibility approach analysis are based on the fuzzy set theory and fuzzy measures.

Possibility theory uses the possibility distribution rather than a single figure, which can be described by incorporating fuzzy membership functions. For example, if the failure rate of pipes is fixed then its risk may be a single number. However, if there are variations in the pipe's condition, in pipe location, or in other factors that contribute to pipe failure, a possibility distribution is useful for analyzing the risk. For each uncertain element, x in a set X, $\pi(x)$ expresses the degree of possibility of x. When $\pi(x) = 0$ it means that the outcome x is considered an impossible situation, whereas when $\pi(x) = 1$ it means that the outcome x is possible (Dubois, 2006). The possibility function gives rise to probability bounds, upper and lower probabilities, referred to the necessity (Nec) and possibility measures (Pos).

The *possibility* of an event A, $Pos(A)$ and the necessity measure $Nec(A)$ are defined by

$$Pos(A) = \sup_{(x \in A)} \pi(x) \tag{2.14}$$

$$Nec = 1 - Pos(not\ A) \tag{2.15}$$

If P is a family of probability distributions for all events A, $Nec(A) \leq P(A) \leq Pos(A)$ then $Nec=inf$ $P(A)$ and $Pos(A)= sup\ P(A)$; where, inf and sup are with respect to all probability measures in P. Thus, possibility theory is similar to probability theory because it is based on the set functions. It differs by the use of a pair of dual set functions - which are possibility and necessity measures.

For example, if x is an uncertain parameter with its values in the range of $[1, 3]$ and a likely value of 2, then a triangular possibility distribution will be represented as in Figure 2.9 (a). From this possibility function, if we define alpha-cut sets $F_\alpha = \{x : \pi(x) \geq \alpha\}$, for $0 \leq \alpha \leq 1$ such that $F_{0.50} = [1.5, 2.5]$ is the set of x values for which the possibility function is greater than or equal to 0.5. From the triangular possibility distribution in Figure 2.9 (b), if A expresses that the parameter lies in the interval $[1.5, 2.5]$, then $0.5 \leq P(A) \leq 1$. Thus, the possibility indicates values that are possible and necessity signifies values that are required to be satisfied. Hence, the necessity measure is interpreted as a lower level for the probability, and the possibility measure is interpreted as an upper limit. Using subjective probabilities, the bounds reflect that the analyst is not able to precisely assign the probability (Dubois, 1998).

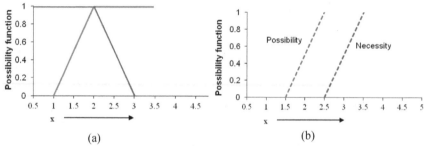

(a)　　　　　　　　　　　　　(b)

Figure 2.9 (a) Possibility function for a parameter on the interval $[1, 3]$, with maximum value at 2

(b) Bounds for the probability measures based on the possibility function in (a).

The deduced cumulative necessity measures $Nec = (\infty, x]$ and possibility measures $Pos = (-\infty, x]$ are shown in Figure 2.9 (b). These measures correspond to the lower and upper cumulative probability distributions for the uncertain parameter x. Hence the bounds for the interval $[1, 2]$ is $0 \leq P(A) \leq 1$. These bounds can be interpreted as the interval probabilities. The interval bounds are those bounds beyond which an analyst is not able to precisely assign a probability given the current information.

It has to be noted that possibility theory has been synonymously used as the fuzzy set theory. The degree of belief assigned to the possibility distribution function is equivalent to membership function in fuzzy number (Zadeh, 1978; Dubois, 2006). Further detailed analysis on possibility theory are also available in Dubios (2006); Helton et al., (2004); Klir and Smith (2001). The application of possibility theory is considered useful to analyse uncertain and imprecise nature of information. Some of the limitations discussed in the possibility theory include:

- Determination of possibility and necessity measures is required during the uncertainty prediction (Dubois and Prade, 1998). This is often associated with additional information, which is subjective in nature and often imprecise as well as difficult to obtain.

- Possibility theory determines the possibility of some uncertain parameters in a set through fuzzy membership functions. Constructing these fuzzy membership functions often requires subjective decisions about the shape and limits of the parameter values with non-zero membership. This introduces an additional source of subjective uncertainty into the analysis.

2.3.4 Hybrid approach to uncertainty analysis

Risks in UWS system can be induced due to multiple sources of hazards, such as droughts, flood, wilful attack, and physical failures. In most of the cases, each hazard affecting system performance has different type of uncertainty. A hybrid approach to uncertainty analysis considers both probability and fuzzy based methods in a single framework.

Ferson and Ginzburg (1996) discussed the need for distinct methods of uncertainty analysis to adequately represent the random variability and imprecision in risk assessment. Kaufaman and Gupta (1985) introduced hybrid numbers which express randomness and imprecision and developed special arithmetic for the propagation of uncertainty in decision-making. Cooper et al. (1996) extended the work of Kaufman and Gupta (1985) to risk assessment. Guyonnet et al. (2003) develop a hybrid approach for addressing uncertainty in risk assessment. This hybrid approach propagates randomness separately using Monte Carlo and impreciseness by the fuzzy interval analysis. The resulting fuzzy random variable is further analysed with special procedures. This method, particularly output analysis, was further improved by other researchers, such as (Baudrit et al., 2006; Baudrit et al., 2007).

Main objective of applying hybrid approach was the recognition of importance of data used for the risk assessment or any decision making process such that no information is added or lost during the analysis. To achieve this, probability theory and fuzzy set theory may be used separately or in combination to successfully strengthen and complement each other. Moving from one theory to another as appropriate facilitates the utilization of the advantages from both theories (Klir, 1987; Klir and Yuan, 1995; Dubois and Guyonnet, 2011). This is achieved by: (i) converting all the information in one form and propagate it homogenously through the models, and (ii) describing all the information separately and propagating them heterogeneously. Any one of the options is possible for representation and propagation of uncertainty in a complex system. For the first case, all the information should be transformed into a common framework (either probability or fuzzy) whereas the second one needs complex analysis techniques that require higher computational efforts for a complex system. This thesis applies the first approach of uncertainty analysis that has been discussed in Chapter 3.

2.4   Proposed Risk Assessment Framework

The review on risk assessment and uncertainty analysis discussed in the previous sections has clearly shown that the existing risk assessment frameworks are unable to adequately address the multiple sources of uncertainty that exist in a complex and socio-technical system like UWS. For example, while analysing a level of services of water supply system, the input parameters used for the modelling the systems are pipe characteristics (e.g., pipe size, length, age etc.) and nodal demand. The nodal demand depends on leakage rate, which is difficult to predict precisely. The imprecision also exists on pipe roughness, which will be affected by changes in the physical and environmental condition of the buried pipes. In this case, nodal demand can be represented in a probabilistic form whereas the pipe roughness will be imprecise and best described by fuzzy membership functions. Similarly, in risk modelling, the likelihood of failure of a system's performance is often analysed by existing systems models, which generate the systems performance in a probabilistic form (e.g., flood depth). However, the consequences of the risks (i.e., losses and damages from the flood) are very difficult to model probabilistically because they are complex, imprecise, and qualitative. In order to address those challenges, this research purposes a new risk assessment framework based on hybrid approach to risk and uncertainty analysis as shown in Figure 2.10. The proposed risk assessment framework involves a number of key steps, including

i)   Setting of the objective of risk analysis and system model,

ii)   Hybrid approach to uncertainty analysis,

iii)   Fuzzy set theory based risk assessment method, and

iv)   Fuzzy set theory based multi-criteria analysis for risk management.

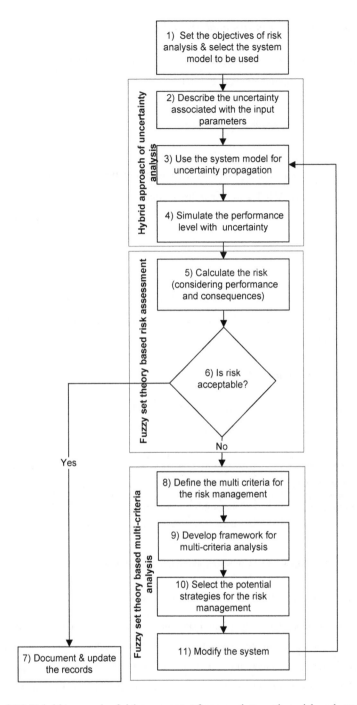

Figure 2.10 Hybrid approach of risk assessment framework to analyse risk and uncertainty for strategic decision making in UWS

i)   Setting of the objective of risk analysis and system model

The first step of analysis is to set the objectives of risk analysis. The objectives of analysis could be, for example, selection of potential alternative water sources for the future water scarcity, controlling the emerging pollutants in the systems, reducing the potential water system failure, ensuring the level of services by marinating the systems performances etc.

Based on the set objective, the system model is applied for the performance analysis. The system model could be newly developed or a commonly used existing UWS model. Some of the commonly used UWS model include, EPANET 2 for water distribution systems, SWMM and MOUSE for urban drainage, AQUACYCLE and UVQ for integrated UWS modelling, STOAT for WWTP simulation, MIKE 11 that simulates flow and water level, water quality and sediment transport in rivers, and City Water of the SWITCH for urban water modelling (see detail in (Mitchell et al., 2001; Mitchell and Diaper, 2006). This step also describes the relationships among the assessment and measurement endpoints, the data required, and the methodologies that will be used to analyze the data. It is noted that the analysis can also be based on the facts based or literature review that ensure the sufficient information for the analysis.

ii)   A hybrid approach to uncertainty analysis

This research proposes a hybrid approach to uncertainty analysis. A general process of uncertainty analysis based on steps: 2, 3 and 4 of the framework shown in Figure 2.10 is presented in Figure 2.11.

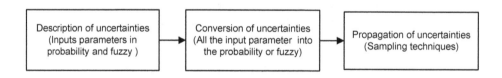

Figure 2.11 A hybrid approach to uncertainty analysis

Major steps in the hybrid approach to uncertainty analysis include:

- *Description of input information* based on data types: If the available information is of a quantitative nature—as in, for example, the per capita water consumption rate in a city, intensity of rainfall, temperature variation etc.—then this input will be described by a probability distribution function. If the available information is subject to impreciseness—as in pipe roughness, quality of groundwater, public perception on quality of services available etc. —a fuzzy memberships function is used for representing the information.

- *Conversion of input information* from one form to another (probability to fuzzy or vice verse) using the algorithms as proposed in Chapter 3. It should be noted that the goal is to convert all of the input parameters into one form so that it can be propagated homogenously through the analysis model.

- *Propagation of uncertainty* is undertaken using simulation techniques such as Monte Carlo, Latin hypercube sampling and fuzzy alpha cut technique as discussed earlier in this chapter and further discussed in Chapter 3.

- The *output* is presented in a probability distribution function or fuzzy membership function and statistical properties are derived.

A critical review on the hybrid approach to uncertainty analysis, proposed methodology, and necessary algorithms are presented and discussed with illustrations in Chapter 3.

iii) Fuzzy set theory based risk assessment

Steps 5 and 6 in the proposed framework are the risk assessment. As discussed, risk in a system is the combination of likelihood of the failure of performances and the consequences of failure (i.e., severity of the failure). For example, in a water supply system, drought, flood, system failures, or wilful incident, etc. could have negative impacts on UWS performances. The consequences or the severity of the failure measures the potential impact to the economic, people and environment due to the systems performances failure. Most of the time, the likelihood of failure in the system is calculated using a probabilistic approach. However, consequences are difficult to quantify. This research applies the quantitative approach of risk assessment based on the fuzzy set theory. The main motivations for using the fuzzy set theory are presented below.Further detail representation and discussion in likelihood of risk, severity of risk and risk calculation based on the fuzzy set theory are presented and discussed in Chapters 3 and 4.

- Risk is not absolutely objective in nature, but rather relative and subjective. It is usually a fuzzy concept in that there does not exist a unique risk associated with a hazardous event occurring in a given period. Therefore, risk assessment deals with quantities which are inherently imprecise and whose future values are uncertain. Usually, linguistic categories or levels (e.g., very high, high, medium, low, very low, etc), instead of absolute values, are adopted. This is because each linguistic category can deal with various and uncertain risk values by including a range of numbers.

- Quantitative risk assessment builds on the existence of probabilities. In general, probabilities are derived from historical records, statistical analysis, and/or systemic observations and experimentation. Often, however, situations arise in which pre-existing data is so sparse and experimentation is so impractical that probabilities must be supplemented with subjective probabilities or based on expert judgment.

- The advantage of the frequency approach is that it can be applied to any situation that can be repeatedly observed. Unfortunately, many events cannot be measured this way simply because they are rare or cannot be repeated a sufficient numbers of times to allow a meaningful determination of relative frequency. For example, one can estimate future probability of flood fatalities in a given area based upon historical data for that area only if future conditions remain unchanged.

- The real meaning of risk in practice varies and is application-specific. Risks are measured in different units; and even similar risk values may indicate different levels of influence in different applications. Furthermore, risks measured in different units are difficult to compare or aggregate in a risk assessment. As an alternative to the absolute numerical values, risk analysts, system managers, policy makers, and general users can more easily use risk levels described in relative measures.

- In certain circumstances, it is not unusual for analysts to be *more* confident in giving risk evaluations in terms of risk levels rather than numerical values. Risk of human wilful attack on UWS assets is one example.

iv) *Fuzzy set theory based multi-criteria analysis for risk management and decision-making*

Risk management involves a decision-making process for selecting an option or strategy to control or reduce the calculated risk. Risk management is thus an analytic process to enable decision-makers to address the identified risks, particularly answering the key questions such as what can be done, what options are available, and what are the benefits and costs of each option. Thus, risk management suggests risk-reducing measures.

Considering the complexity and multidimensional nature of UWS, this research proposes multi-criteria based risk management approach. The multi-criteria analysis allows the selection of an option for risk management by assessing different criteria in different dimensions of a system (e.g., environmental, economic, social and technical). Detailed discussions on the fuzzy set theory based multi-criteria analysis technique is presented in Chapter 4.

2.5   Discussion and Conclusions

This Chapter presented a state of the art review on risk assessment, risk management, risk analysis approaches, and frameworks. The review established many definitions, relations, and approaches for the risk and uncertainty analysis for a decision-making process. The main outcome of this Chapter is a new risk assessment framework for strategic decision making in UWM.

There is no standard definition of risk. Risk is a term often used interchangeably with other terms like chance, likelihood, and probability to indicate that there is uncertainty about the state of something under discussion. Most meanings carry negative connotations of danger, peril, or loss. As such, a risk is described in terms of the minimum required level of system performance. The system performances will include minimum levels of services for water quality and quantity, system failure rate, flood depth, etc. The severity measures the numbers of peopled affected due to performance failure in a system.

Uncertainty describes the quality of knowledge concerning risk that needs to be faithfully represented, propagated, and presented for decision-making. Uncertainty stems from a number of sources, such as the natural variability of a system or phenomenon, the multiple levels of social interaction in built systems, and the technological complexities in UWS. Generally, uncertainty can be reduced with an increased state of knowledge or the collection of more data. Sources of uncertainty refer to the location and causes that give rise to uncertainties. The unknown information may be stochastic, random, vague, or subjective. For simplicity, this thesis considers only two types of uncertainty: aleatory and epistemic. Aleatory uncertainty represents inherent variability and therefore cannot be reduced by the collection of more information or data. Epistemic uncertainty is due to subjectivity, ignorance, or a lack of information in any phase of the modelling process.

The proposed risk assessment framework is new as it incorporates the hybrid approach to uncertainty analysis, which was not adequately addressed in the majority of the existing risk assessment frameworks. The hybrid approach to uncertainty analysis aims to address different types of uncertainty using different theories at the various stages of a problem solving process. To maximize the advantages of the various theories, we must move from one theory to another as appropriate (Klir, 1987; Klir and Yuan, 1995; Dubois and Guyonnet, 2011). This requires the theories be integrated such that the results obtained by one theory are converted—using justifiable transformations—into equivalent representations of uncertainty in another theory. At the same time, during any risk assessment or decision making process no information should be added or lost. To achieve this objective, this research develops an evidence theory based conversion technique. Further detailed analysis and discussion on the hybrid approach is presented in Chapter 3.

The risk in a system is analysed using a fuzzy set theory based semi-quantitative technique. It is different from other existing methods as it allows the utilization of the strengths of advanced systems models to assess the performances and obtain the results in quantitative/probabilistic form. At the same time, it recognises the importance of risk aversion, risk perception and the complexities of the consequences of modelling in UWS as "perceived risks" are different from "real risks", and risk perception depends on a number of subjective factors. Therefore, risk-based decision-making will vary according to the perception of the identified risks and the capacity for risk aversion within a community. Commonly, they are captured in qualitative forms from expert views. Therefore, the next important characteristic of this framework is its ability to capture the majority of the uncertainty during the decision making process. The detail of the proposed approach for risk assessment is presented in Chapter 3.

Next important feature of the framework is a multi-criteria based decision-making framework for risk management and UWM. The MCDM framework is fuzzy based and unique as it allows to capture multiple-criteria in economic, social and technical dimensions and information contained by data (i.e., crisp, quantitative or qualitative). The framework is hierarchical to address the complexities of UWS. It allows analysing the performances of a component of a system or system or systems at different levels. The framework also includes the Analytical Hierarchical Programming (AHP) to assign the relative importance of the criteria and indicators. This also recognises the multi sectors involvement in UWM and it allows interact the stakeholders, which was not possible in most of the risk assessment framework. Detail discussion on multi-criteria analysis of the decision-making framework is presented in Chapter 4.

Chapter 3   A Hybrid Approach of Uncertainty Analysis and Uncertainty Modelling

The objective of this Chapter is to present a review of hybrid approaches to uncertainty analysis and modelling, develop a new hybrid approach of uncertainty modelling, and demonstrate the newly developed approach in a real case analysis. The Chapter begins by discussing the different approaches of uncertainty description and propagation in the decision-making process. It then presents a critical review on existing methods of hybrid approaches of uncertainty analysis. Based on the review, a new hybrid approach of uncertainty analysis is developed. This hybrid approach is supplemented by modelling algorithms for MCS, LHS, Boot Strapping technique, and fuzzy alpha-cut technique. The Chapter finishes by presenting an application of the proposed hybrid approach of uncertainty analysis and risk assessment technique for future water availability analysis in Birmingham, UK in 2035. A part of this Chapter was also presented in Khatri and Vairavamoorthy (2009); Khatri and Vairavamoorthy (2011a) Khatri and Vairavamoorthy (2011b); Khatri and Vairavamoorthy (2013b).

3.1   Introduction

As discussed in Chapter 2, there are different theories and methods for uncertainty analysis. Some of the methods are using probability theory, such as analytical methods (Tung, 1996), sampling methods (Kuczera and Parent, 1998), generalised likelihood method (Beven and Binley, 1992) and model error methods (Montanari and Brath, 2004). Other widely used methods are based on fuzzy set theory including fuzzy extension principles (Zadeh, 1975) and fuzzy alpha-cut technique (Maskey et al., 2004). Similarly, there are a few methods that are based on hybrid approaches including hybrid-number for imprecision and randomness (Kaufmann and Gupta, 1985) and hybrid propagation methods (Guyonnet et al., 2003).

Generally, the sources and types of uncertainty define the selection of theories and methods to be applied in uncertainty analysis. Thus, the first step of the uncertainty analysis is the *description or representation of uncertainty* based on the information contained by the input parameters. For example, if a given parameter is deterministic and known, it will be represented in a *crisp* form with certainty, such as pipe length and diameter. If a parameter is variable with sufficient precise measurements, it will be represented by *probability distribution* functions, such as rainfall and runoff in a catchment. If the information is qualitative and imprecise, such as quality of services, deterioration rate of underground pipes, it will be represented by *fuzzy membership function* or *possibility distribution*. Figure 3.1 presents a flow chart developed by Dubois and Guyonnet (2011) that illustrates how uncertain information can be described/represented based on the information that is also relevant to UWS.

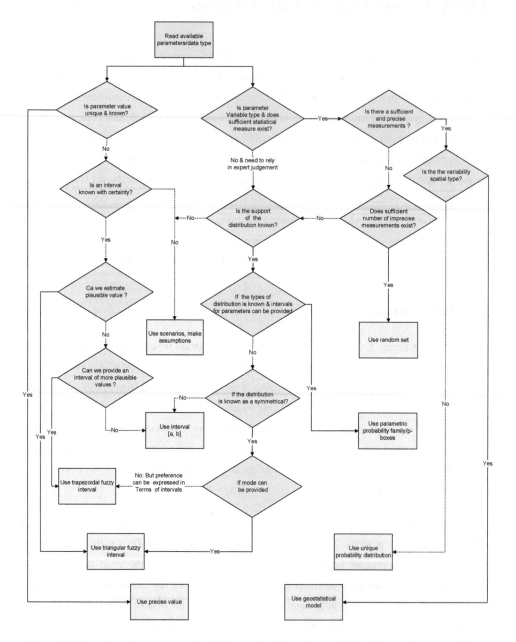

Figure 3.1 Flow chart for representation of uncertainty based on available information adopted from Dubois and Guyonnet (2011).

Moreover, the chart shows the appropriate representation for a number of other types of data including deterministic quantities; known constant values; interval values; insufficient information for defining a parameter interval; a random variable; and temporal variability with sufficient data for a probability distribution. The proposed flowchart brings the user to realize that there is no one-size-fits-all method for describing/representing uncertainty. As listed, uncertain parameters could be described by interval analysis, fuzzy membership function, or probability distribution functions based on the information, which is generated by either statistical analysis of data or by expert views.

In the case of probabilistic data, the probability distribution models may be empirical, parametric, or combinations of both. A parametric probability distribution model is described by parameters that can describe a large number of data points in a compact manner, such as a normal distribution, lognormal distribution, uniform distribution etc. Statistical goodness-of-fit tests provide a quantitative measure of the assumed probability distributions, but may only apply to parametric distributions. Three commonly used goodness-of-fit tests for parametric distributions include the chi-square test, the Kolmogorov-Smirnov (K-S) test, and the Anderson-Darling (A-D) test. More detailed discussion on distribution selection criteria can be found in Ang and Tang (1984); Morgan and Henrion (1990); Ross (2009); Cullen and Frey (1999), among others. In the case of fuzzy memberships, the shape and numbers of the uncertain parameters are selected by using the experts' knowledge. The bias and incompetency need to be treated with the great care while selecting a shape of the membership functions (Zedeh, 1989; Klir and Yuan, 1995).

Propagation of uncertainty through an analysis model is the next important step in risk assessment and decision making processes. Uncertainty propagation methods can be propagation of distributions, or propagation of moments. The first order second moment method is an example of the propagation of moment, whereas Monte Carlo Simulation is an example of the propagation of probability distributions. There are numerous methods of propagation in literature and in some cases, it can be unclear which method to apply in a given condition. It has been reported that most analysis processes in real cases are complex, and clear guidance for their understanding and application is lacking (Pappenberger and Beven, 2006). Pappenberger et al. (2005) presented a decision tree for selecting an appropriate method of uncertainty analysis for a given situation as shown in Figure 3.2. As presented, a method of uncertainty propagation depends on the type of data, complexities of models and model output. Further detailed discussion on the flow chart is available in Pappenberger et al. (2005).

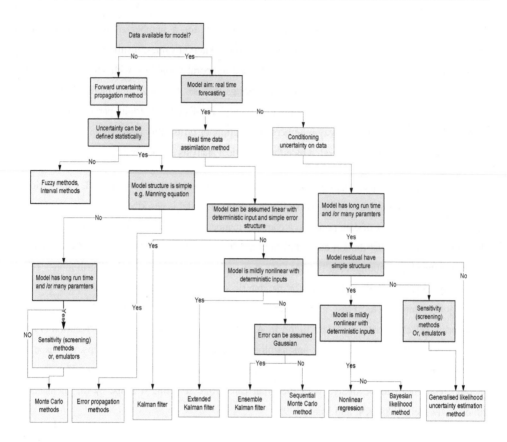

Figure 3.2 Decision tree for uncertainty analysis (Pappenberger et al., 2005)

There is not any clear guidance for the selection of any particular method of uncertainty analysis that produce the best results for a specific problem and analysis. Selection of any uncertainty analysis method depends on many factors including the types of information available, complexities of a model and purpose of the analysis —if this is decision analysis or forecasting for the future. UWS is a complex entity; modelling of UWS requires multiple types of input parameters, for example:

- Deterministic types of data of physical properties of the infrastructure systems, such as length, diameter, thickness, other physical dimensions of infrastructure systems, etc.

- Statistical and probabilistic types of data, such as historical temperature, rainfall and runoff data, population growth, surface water discharge, water demand, rate of wastewater generated, rate of system failure, water quality variation, runoff rainfall relation, etc.

- Imprecise and qualitative types of data, such as ground water level, ground water pollution, infiltration rate, roughness of pipes, public behaviour, water leakage and losses, willingness to pay for the services, etc.

- Unknown and ignorance, such as limited data availability about the system's physical characteristics, magnitude and intensity of climatic induced events such as future extreme rainfall, droughts, other natural disaster such as earthquake, wishful attack on infrastructure systems, rate of groundwater pollution, etc.

UWS modelling and analysis will require to deal with the multiple sources and types of uncertainty. This implies that for the faithful representation of available information in risk assessment or any decision making process in UWS requires application of more than one theory for uncertainty analysis.

3.2   Existing Methods for Hybrid Approach to Uncertainty Analysis

As discussed, the hybrid approach to uncertainty modelling allows us to address both types of uncertainty information in a single framework. This could be achieved either by transforming all the information into one form and propagating homogenously through a model (i.e., either random simulation or fuzzy approach), or by describing all the information separately and propagating heterogeneously. Therefore, transformation and propagation are the main process in a hybrid approach of uncertainty analysis. The next section reviews and discusses the principles and techniques used for a transformation and propagation of uncertainty.

3.2.1 Hybrid approach by homogenous propagation

Principles governing transformation

There are several methods for the conversion/transformation of one form of uncertain information into another. Most transformation techniques are based on several principles such as Zadeh consistency principle (Zadeh, 1978), Dubois and Prade's consistency and preference preservation principles (Dubois and Prade, 1982), Delgado and Moral's maximal specificity principle (Delgado and Moral, 1987), Geer and Klir's information-preservation principle (Geer and Klir, 1992).

Zadeh (1978) consistency is a set of statements hypothesizing a relation between possibility and probability such that a transformation of possibilistic data to probabilistic data, and vice versa, will be possible in a consistent manner. This principle presents the *degree of consistency* between possibility and probability distributions for transformation from possibility into probability. Let $U = \{x_1, \dots x_n\}$ be a finite set of discrete events, X is variable taking a value in U ; $p(x_i)$ and $\pi(x_i)$ are probability and possibility that $X = x(i)$ respectively. According to Zadeh's principle, if an event is impossible, it is bound to be improbable, which can be represented by

$$\pi(x) = 0 \Rightarrow p(x) = 0 \ \forall x \subseteq X \ and \ \pi(x_i) > \pi(x_j) \Rightarrow p(x_i) > p(x_j)$$

(3.1)

where $\Rightarrow$ means implication.

The measure of consistency between $\pi(:)$ and $p(:)$ is given by

$$\gamma = \sum_{x_i = U}^{n} \pi(x_i).p(x_i)$$

(3.2)

Thus, greater γ because the greater the consistency between the two event such that if $\pi(x) = 0 \ \forall x$ then $\gamma = 0$, no consistency is available.

As shown, if an event is impossible, it is bound to be improbable. Zadeh (1978) notes that the probability-possibility consistency is not a precise law or a relationship between possibility and probability distributions. Rather, it is an approximate formalization of a heuristic connection. It has no logically sound or empirical support behind that a lessening of the possibility of an event tends to lessen its probability, but not vice-versa. Dubois and Prade (1982) point out that Zadeh's possibility expresses the degree of ease and is different from epistemic possibility, which expresses the uncertainty of occurrence.

Dubois and Prade (1982) pointed that the transformation between probability and possibility will be governed by the *principles of consistency* (Equation 3.3) and *preference preservation* (Equation 3.4). Possibility-probability consistency principles imply that one should select a possibility distribution (π) with probabilities data (p), such that the transformation of possibilistic data to probabilistic data and vice versa will be possible in some consistent manner. This implies that the transformation from probability into possibility should be guided by preference preservation principles so that most of information could be preserved.

$$\prod(A) \geq P(A)$$

(3.3)

$$p(x_i) > p(x_j) \Leftrightarrow \pi(x_i) > \pi(x_j)$$

(3.4)

where $\Leftrightarrow$ means equivalence, and $P(A)$ and $\prod(A)$ $(A \subseteq U)$ are probability and possibility measures defined by $P(A) = \sum_{x_i \in A} p(x_i)$ and $\prod(A) = \vee_{x_i \in A} \pi(x_i)$, respectively (here, $\vee$ means maximum).

The possibility into probability transformation is guided by the principle of insufficient reason, which aims at finding the possibility distribution that contains as much as uncertainty as possible while retains the features of possibility distribution. The transformation proposed by Dubios and Prade is possible in both directions and satisfies the consistency and the preference preservation principles. However, this is only applicable for unimodal cases and there is no guarantee that this is the only one and final form of transformation using those principles (Yamada, 2001).

Delgado and Moral (1987) envisages that the information included in a possibility distribution is less than that in probability. Therefore, the possibility distribution generated from probability should include fuzzy sets only minimally, so that the possibility distribution is maximally specific. It means, the possibility distribution should be selected in such a way that it can be must informative while transformation. He provided the relation to satisfy the consistency principle (Delgado and Moral, 1987). Assuming that both a possibility and probability distribution are given, and that the possibility satisfies $\pi(x_i) \geq ... \geq \pi(x_n)$. In this case, the consistency principle is satisfied, if and only if (iff) the following conditions hold (Delgado & Moral, 1987):

$$\pi(x_i) \geq \sum_{j=i,n} p(x_j), \forall x_i \in U \tag{3.5}$$

when a probability distribution satisfying $p(u_i) \geq ... \geq p(u_n)$.

Klir (1990b) emphasises the unique connection between uncertainty and information and proposes a *principle of information invariance or information preservation.* The principle argues that the transformation of uncertainty from one theory into an equivalent representation for another theory should satisfy at least the following two requirements: (i) the amount of uncertainty inherent in the situation of concern be preserved when we move from one theory into another; and (ii) all relevant numerical values in one theory be converted to their counterparts in another theory by an appropriate scale. Klir added that to guarantee the first requirement no uncertainty is unwittingly added or eliminated solely by changing the mathematical theory by which a particular phenomenon is formalized. If the amount of uncertainty were not preserved, then either some information not supported by the evidence would unwittingly be added by the transformation (information bias) or some useful information contained in the evidence would unwittingly be eliminated (information waste). In either case, the model obtained by the transformation could hardly be viewed as complementary to its original.

The preservation of scale which may be in any form such as ratio scale: $\pi(x_i) = \delta * p(x_i)$; interval scale $\pi(x_i) = \delta * p(x_i) + \beta$; difference scale: $\pi(x_i) = p(x_i) + \beta$; log-interval scale: $\pi(x_i) = \beta * p(x_i)^{\delta}$; ordinal scale: $\pi(x_i) = \pi(x_j) \Leftrightarrow p(x_i) = p(x_j)$ and $\pi(x_i) > \pi(x_j) \Leftrightarrow p(x_i) > p(x_j)$.

Geer & Klir (1992) show that ratio scale transformation cannot preserve uncertainty, since there is solely one variable δ, which is completely determined by the normalisation condition. Usually, scale preservation ensures ignorance and preference preservation (Oussalah, 1999). Geer & Klir propose the log-interval scale transformation to satisfy the assertion in both directions, given by

$$\pi(x_i) = \left\{ \frac{p(x_i)}{p(x_1)} \right\}^{\delta} \tag{3.6}$$

where x_1 is an element having the largest probability, and δ is factor and see in Geer & Klir (1992)

Based on the aforementioned principles several techniques of transformation have been developed. Most of these principles suggest ensuring information preservation while transforming. Some of the methods developed based on principles presented earlier are presented in the next section.

Techniques for transformation

Zadeh (1978); Dubois and Prade (1982); Delgado and Moral (1987); Geer and Klir (1992) developed techniques of transformation with their own principles of transformation as presented in earlier section. Others have developed several techniques of transformation based on their principles and empirical methods. For example, Gupta (1993) develops a transformation method that converts fuzzy data (possibilistic information) into a normal probability distribution (probabilistic information) using empirical method. This method is applicable in cases where fuzzy restrictions on estimates are expressed in terms of symmetrical fuzzy numbers (i.e., expressed as $(x_o \pm s)$, is symmetrical around x_o where $\mu_x(x_o) = 1$ and s is known as spread). The purpose of this distribution is to preserve symmetry. Here, the spread 's' may determine the degree of variation. Using the symmetrical fuzzy numbers of a variable, the mean $(= x_o)$ and the variance s of the probability distribution can be determined from x_o and s respectively. This technique is very simple but is only applicable, if the membership functions are symmetrical.

Jumarie (1994) proposes a possibility-probability transformation technique using deterministic functions, obeying the information invariance principle. Yamada (2001) proposes an evidence theory based approach for the transformation of possibility information into probability for discrete cases only. Dubois et al. (2001) provide a new approach based on the semantics of the transferable belief model, which preserves the most informative approximation of a probability measure. Similarly, Dubois et al. (2004) investigate the new properties of transformations by means of probability inequalities and propose a transformation in the unimodal and continuous case.

Anoop et al. (2006) proposes technique of conversion of probabilistic information into a certain fuzzy set for seismic hazard analysis in terms of least-square curve fitting, entropy function, and Hausdorff distance measure between the probability and fuzzy sets. This method shows that the conversion of uncertainty is equivalently obtained by preserving the elements of the fuzzy set and their representations. However, this approach is applicable only for triangular and trapezoidal fuzzy membership functions.

Recently, there has been further development on the transformation of one form of information into the other, especially for continuous distributions and multi-modal cases. Florea et al. (2008) have proposed a random set theory based unified framework in which probability theory, evidence theory, possibility theory and fuzzy set theory can be represented. The paper also proposed approximation techniques where large numbers of focal elements need fusion. Mauris (2010) extends the previous work on unimodal distributions by proposing a possibility representation of bimodal probability distributions based on the identification of level cuts of the probability density to the alpha-cuts of the possibility distribution. Serrurier and Prade (2011) propose a likelihood function for possibility distributions for continuous and discrete cases. They argued that given a set of data following a probability distribution, the optimal possibility distribution with respect to likelihood function is the distribution obtained as the result of the probability-possibility transformation that obeys the maximal specificity principle. The paper also showed that when the optimal distribution is not available, a direct application of this possibilistic likelihood provides more faithful results than approximating the probability distribution and then applying the probability possibility transformation. Those works have contributed to understanding the relation between possibility and probability. However, Florea et al. (2008) and Mauris (2010) works are very difficult to apply in complex modelling and the Serrurier and Prade (2011) approach does not take into account the quantity of data available.

The reviews signify that this area of research is still in an infancy state. Most of the techniques considering discrete cases are not applicable for continuous cases; several techniques are analysing the unimodal; other techniques address bimodal distributions types; many are not related with the fuzzy membership function and probability distribution function. Several of the techniques are very complex to apply directly; few are only case specific solutions that necessitate further analysis for their application. Further detailed discussion on transformation techniques are available elsewhere (Geer and Klir, 1992; Oussalah, 1999; Dubois et al., 2004; Dubois, 2010; Sadeghi et al., 2010; Ali and Dutta, 2012; Dhar, 2012). This research proposes a new simplified transformation technique based on evidence theory, which is presented in the section 3.3.

3.2.2 Hybrid approach by heterogeneously propagation

An alternative strategy for dealing with complex systems that have different sources and types of uncertainty is by describing all the information separately and then propagating them heterogeneously. This will require joint propagation techniques of probabilistic and fuzzy information in a single framework.

In heterogeneous framework, many researchers have suggested hybrid methods for propagating the probabilistic information and fuzzy numbers in the same framework (Cooper et al., 1996; Ferson and Ginzburg, 1996; Tonon et al., 2001; Guyonnet et al., 2003; Dubois, 2010). In the early development, Kaufmann and Gupta (1985) suggested that instead of confounding the two forms of uncertainty they could be considered together but separately in a pair (μ, p) where μ is a fuzzy number and p is a probability distribution. These pairs, which they termed "hybrid numbers," can be added together by convolving the respective elements according to normal rules for fuzzy arithmetic and probability theory. Their formulation of hybrid numbers allows for addition and subtraction. Although they developed a new concept, their formulation of hybrid numbers does not directly allow full hybrid arithmetic (e.g., the product of a completely fuzzy number and a completely probabilistic number is undefined) where the plus signs on the right side of the equation represent fuzzy max-min convolution and ordinary probabilistic sum-product convolution (Cooper et al., 1996).

Ferson and Ginzburg (1996) have also emphasized the need for different methods to propagate ignorance and variability. They suggested the use of interval analysis to propagate ignorance, and probability theory to propagate variability. They also showed how ignorance and variability can be represented simultaneously and manipulated in a coherent analysis that does not confound the two forms of uncertainty and distinguishes what is known from what is assumed. Cooper et al. (1996) further extended the work of Kaufmann and Gupta (1985) to describe other formulations for hybrid numbers for multiplication, division, and other operations. They propose that hybrid numbers can be thought of in two ways: a *fuzzy* probability distribution, or a random distribution of *fuzzy* numbers. They showed that two concepts are equivalent, but have complementary interpretations and calculation strategies. Although logically sound, the technique could not be used widely in risk analysis because of computational complexity.

With the same principle of Ferson and Ginzburg (1996), Guyonnet et al. (2003) proposed a joint propagation technique in the name of hybrid propagation. Guyonnet et al.'s (2003) hybrid propagation method combines random probabilistic variables with fuzzy variables. It uses repeated Monte Carlo sampling to process the uncertainty associated with probabilistic variables, and uses fuzzy interval analysis (fuzzy alpha-cuts) to process the uncertainty related to fuzzy variables. Based on this technique, Baudrit and Dubois (2005) analysed the different methods (i.e., independent random set approach, casting the "hybrid" approach in the random set setting, conservative random sets approach, dependency bounds convolution approach) of propagation for joint objective and subjective uncertainty propagation with an example in a risk assessment. They concluded that the result of the hybrid approach was conservative.

This approach has been further analysed by Baudrit et al. (2006) for the sensitiveness of input parameters (e.g., small uncertainty verses large uncertainty for both the random and imprecise parameters) and suggested the modified technique for post processing the output results based on the evidence theory. Baudrit et al. (2006) additionally proposed an alternative uncertainty propagation method, called the independent random set method, where the random sampling procedure is applied not only to the probability distributions, but also to the fuzzy intervals. This joint-propagation method was applied to problems of groundwater contamination by Baudrit et al. (2007). The result was compared with the probabilistic approach (using MCS). The result was found between the upper and lower limits of the joint propagation techniques. The benefit highlighted by the paper are: (i) results of the joint-propagation methods are "families" of cumulative probability distributions represented by their upper (plausibility) and lower (belief) limits and the distance between these two distributions are primarily a consequence of the imprecise nature of available information and, to much a lesser extent, of the choice of the propagation method.

Baraldi and Zio (2008) applied the improved hybrid method after improvement by Baudrit et al. (2005) on a case study concerning the uncertainties in the probabilities of occurrence of accident sequences in an event tree analysis of a nuclear power plant and compared the results with pure probabilistic and pure fuzzy approach for uncertainty propagation. Baraldi and Zio found that the cumulative distribution of the accident sequence probability obtained by the pure probabilistic method was within the believe and plausibility functions. The believe and plausibility functions obtained by the fuzzy approach were further away from each other in all cases than the hybrid Monte Carlo and possibilistic approaches. The benefit of using the hybrid approach was the estimation of upper and lower cumulative distributions of the probabilities of occurrence of the accident sequences.

Despite the many improvements in the hybrid methods of uncertainty analysis, it has been applied only in simple cases with few input parameters. This could be due to the complexities in analysis of post processing of the hybrid results and in many cases requiring a different mathematical operation for the analysis. In addition, the technique proposed by Guyonnet et al. (2003) cannot be applied for a case of fuzzy membership function (possibility distribution) which is not normal as the values that do not actually belong to an alpha-cut may be considered in the alpha-cut (Sadeghi et al., 2010). The alpha-cuts of a fuzzy set cannot always be represented by Infimum (*Inf*) and Supremum (*Sup*) values. Moreover, as pointed out by Baudrit et al. (2006) and Sadeghi et al. (2010) this method leads to unrealistic outputs and overestimations. Further discussion on the hybrid approach of uncertainty analysis is available in Dubois and Guyonnet (2011).

3.3 Proposed Hybrid Approach

From the literature review, it is clear that there is much ongoing research in the areas of transformation and the hybrid approach of uncertainty propagation. However, it is very difficult to select a technique that can be readily applied for the transformation of a specific fuzzy membership function into a probability function or vice versa. A similar situation exists for hybrid propagation methods. Theoretically, they are very rich, well discussed, and accepted, but their uses are very limited in real cases, particularly in UWS. This could be due to the complexities in analysis or low sensitivities in the results. Considering these practical limitations, we propose a simplified technique of transformation from one information type into other based on the evidence theory, we recommend the homogenous propagation technique for the uncertainty propagation where either probability theory based sampling techniques or fuzzy extension principle can be employed (Khatri and Vairavamoorthy, 2013b).

3.3.1 Transformation using Dempster-Shafer theory

Dempster-Shafer theory (DST), also known as evidence theory, characterizes the uncertainty caused by partial ignorance, knowledge deficiency or inconsistency about a system by experts (Sentz and Ferson, 2002). The focus of early DST approaches was for the data fusion and artificial intelligence. Recently the DST has been also applied for uncertainty analysis. This research applies the DST for transformation of one form of uncertainty information into the other rather than direct application to uncertainty analysis. Further discussion on DST including uncertainty modelling are available in (Agarwal et al., 2004; Bae et al., 2004; Limbourg and Rocquigny, 2010; Ferdous et al., 2011).

Dempster-Shafer theory

DST characterizes the uncertainty parameters with a definition of frame of discernment. The frame of discernment U is a set of mutually exclusive alternatives, which allows the power set " A " to have a total of $2^{|U|}$ subsets in the domain, where $|U|$ is the cardinality of the frame of discernment. For example, let water scarcity risk be represented by Low, Medium and High risk denoted by {L}, {M}, and {H} respectively with the frame of discernment U={L,M,H}. Its power set comprises 8 subsets (the cardinality is 3), due to closed world assumptions over "union", that is, the possible outcomes are exhaustive and cannot be outside the frame of discernment. The power set A contains the 8 subsets

(i=1 to 8), $A_i = \{1,2,3,...,...,8\}$, i.e., $U(\varnothing)$,{L}, {M},{H},{L,M},{M,H},{L,H}, and {L,M,H}. Thus, depending on the evidence, the probability mass can be described as Low, Medium, High, Low or Medium, Low or High, Medium or High, and Low or Medium or High (the last subset denotes a fully ignorant situation). DST analysis relies on three major measures, namely, basic probability assignment (m or bpa), belief *(Bel)*, and plausibility *(Pl)*.

The basic probability assignment (bpa) is a primitive component of evidence theory. The basic probability assignment (bpa or m) expresses the proportion of all available relevant evidence that supports the claim that a particular element of power set A belongs to the (sub)set A_i but not to a particular subset of A_i (Klir and Yuan, 1995; Klir, 1995). $m(A)$ expresses the degree of belief committed exactly to set A. Every subset A_i for which $m(A_i) \neq 0$, is called a focal element. If F is the set of all focal elements, then the pair (F,m) is called the *body of evidence*. The union of all focal elements of a belief function is called its *core*. The bpa of the null subset $m(\varnothing)$ is 0 and the sum of the basic probability assignments $m(A_i)$ in given evidence set $"< m(A_i), A_i >"$ is "1". Thus,

$$m : 2^U \rightarrow [0,1]; \quad m(\varnothing) = 0; \quad \sum_{A_i \subseteq A} m(A_i) = 1 \qquad (3.7)$$

The mass $m(A_i)$ is defined over the interval $[0, 1]$, but is different from the classical definition of probability. The probability density function is defined in every singleton of U, whereas the basic probability assignment function can be defined on various subset of U.

The lower probability *bound*, belief *(Bel)* for a set A_i is the sum of all the basic probability assignments of the proper subsets A_k of the set of interest A_i, i.e., $A_k \subseteq A_i$. The general relation between bpa and belief can be written as

$$Bel(A_i) = \sum_{A_k \subseteq A_i} m(A_k) \qquad (3.8)$$

$$Bel(\varnothing) = 0; \quad Bel(U) = 1 \qquad (3.9)$$

It should be noted that $Bel(L,M) \geq Bel(L) + Bel(M)$, because DS theory allows some mass to be assigned to less specific subset $Bel(L,M)$ which was not permitted in traditional approaches.

The upper probability bound, plausibility (Pl), is the summation of the basic probability assignment of the sets A_k that intersect with the set of interest A_i, i.e., $A_k \cap A_i \neq \varnothing$.

$$Pl(A_i) = \sum_{A_k \cap A_i \neq \varnothing} m(A_k) \tag{3.10}$$

In addition, the following relationships for belief, doubt (i.e., doubt is the degree of belief that event A does not occur), and plausibility functions hold true in all circumstances:

$$Doubt(A_i) = Bel(\overline{A}); \; Doubt(A_i) = Bel(\overline{A}); \; Pl(A_i) = 1 - Doubt(A_i) \tag{3.11}$$

$$Pl(A_i) = 1 - Bel(\overline{A}_i); \; Bel(A) = 1 - Pl(\overline{A}) \tag{3.12}$$

$$Pl(A) = 1 - Bel(\overline{A}); \; Pl(A_i) \geq Bel(A_i); \; Pl(\varnothing) = 0 \tag{3.13}$$

where, $Pl(U) = 1$

The belief interval (I) is an interval between belief and plausibility representing a range in which true probability may lie. A narrow belief interval represents more precise beliefs and the probability is uniquely determined if $Bel(A_i) = PL(A_i)$ (Yager, 1987). If $I(A_i)$ has an interval $[0, 1]$, it means that no information is available; conversely, if the interval is $[1, 1]$, it means that A_i has been completely confirmed (having full evidence) by $m(A_i)$. Belief functions cannot be directly used for decision-making. It can apply the pignistic transformation. Pignistic probability (Smets, 2000) is a point (crisp) estimate in a belief interval and is given by

$$Bet(A_i) = \sum_{A_k \subseteq A_i} \frac{m(A_k)}{|A_k|} \tag{3.14}$$

The denominator $|A_k|$ in the above Equation represents the cardinality (number of elements) of the (sub) set A_k. The sum of pignistic probabilities is always 1.

The relation between belief, uncertainty, and plausibility is shown in Figure 3.3. Thus, $Pl(A)$ expresses the extent to which one fails to doubt A, or in other words, the extent to which one finds A plausible. The relations between *bpa, belief, and the plausibility function* show that $m(A)$ characterizes the degree of evidence or belief that the element in question belongs to set A alone (i.e., exactly to set A, whereas $Bel(A)$ represents the total evidence or belief that the element belongs to A as well as to the various special subsets of A. Plausibility thus represents the total evidence or belief that the element in question belongs to set A or to any of its subsets, but also the additional evidence or belief associated with sets that overlap with A.

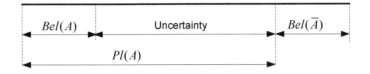

Figure 3.3 Representation of belief, plausibility measures and uncertainty

As discussed earlier, the possibility theory is based on two dual measures, which are special versions of belief measures, and plausibility measures. They are interpreted as lower and upper probability estimates. The probability theory coincides with the sub-area of evidence theory in which belief measures and plausibility measures are equal. Both the probability and possibility measures are uniquely represented by distribution functions, but their normalisation requirements are very different. Values of each probability distribution are required to add to 1, while for possibility distributions the largest values are required to be 1. Both theories are subsumed under a broader theory usually called evidence theory of Demspter -Shafer theory, which is turn is subsumed under a theory referred to as fuzzy measure theory (see Figure 3.4a) (Klir and Parviz, 1992). These special mathematical axioms provide the opportunities for establishing the relationships between possibility (fuzzy measures) and probability distributions. Considering those relations between the different theories of uncertainty analysis, the following section presents the transformation technique.

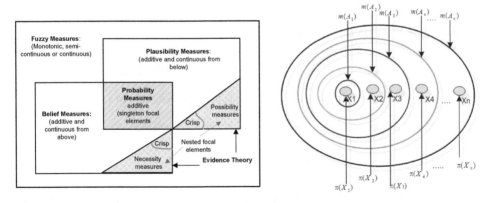

Figure 3.4 (a) Inclusion relationship among belief, plausibility, probability, necessity, and possibility measures adopted from (Klir and Yuan, 1995) (b) Example of a complete sequence of nested subset

Let $U = \{x_1,....x_n\}$ be a finite set of discrete events, X is a variable taking a value in U; $p(x_i)$ and $\pi(x_i)$ are probability and possibility functions. In DST, when all of the focal elements are nested and evidence is *consonant*, $Bel(A)$ and $Pl(A)$ will be equal to the necessity $N(A)$ and possibility measures $\Pi(A)$ respectively, for any $A, B \subseteq U$.

In the nested structure of subsets, the smallest set is included in the next larger set and all of the elements of smaller set are included in the next larger set and so on (Figure 3.4 b). In this case, if $X = \{x_1, x_2, x_3, ..., x_n\}$ is the frame of discernment, the complete sequence of nested subsets of X is given by $A_i = \{x_1, x_2, ..., x_i\}$ for $i = 1, 2, ..., k, ...n$ that implies that $A_1 \subset A_2 \subset A_3 .. \subset A_k .. \subset A_n \equiv X$.

For a case of nested subsets, $A_1 \subset A_2 \subset A_3 ... \subset A_k .. \subset A_n \equiv X$ and necessity and possibility, functions can be expressed in terms of basic probability assignment as:

$$N(A_j) = \sum_{A_k | A_k \subseteq A_j} m(A_k) = \sum_{k=1}^{j} m(A_k) \tag{3.15}$$

$$\Pi(A_j) = \sum_{A_k | A_k \cap A_j \neq 0} m(A_k) = \sum_{k=1}^{n} m(A_k) \tag{3.16}$$

where, k is the arbitrary index set such that $A_k \subset A_j$ and $k \leq j$.

Transformation of possibility and probability

Let $E_p = (F^p, m_p)$ and $E_\pi = (F^\pi, m_\pi)$ are the bodies of evidence defining the probability and possibility distributions, respectively. The transformation between probability $p(x_i)$ and possibility $\pi(x_i)$ can be replaced by the transformation between E_p and E_π, where $m_p(\{x_i\}) = p(x_i)$,
$$F^p = \{\{x_i\} | x_i \in U_1 \cup ... \cup U_{kp}\} \text{ and } F^\pi = \{F_1^\pi, ..., F_{K\pi}^\pi\}.$$

During the conversion process, possibility can be considered as an ordinal scale or ratio scale (Yamada, 2001). In this case, possibility is considered as a ratio scale. It means that values of possibility have their own meaning signifying that if possibility of 0.20 is considered two times more than possibility of 0.10. For this, a principle called *equidistribution* is applied, which was introduced when Dubois and Prade approximated a plausibility measure by a probability measure (Klir, 2006;

Yamada, 2001). When this principle is applied to the transformation from possibility to probability, the value of $m_{\pi}(F_k^{\pi})$ is equally distributed to all elements in F_k^{π}. This is because $m_p(A)$ must be 0 for all A satisfying $|A| > 1$. Using this principle, Yamada (2001) derive a possibility distribution into a probability relation as below.

$$p(x_i) = m_p(\{x_i\}) = \sum_{h=k, K} m_{\pi}(F_h^{\pi}) / |F_h^{\pi}|$$

$$\forall x_i \in G_k^{\pi} = F_k^{\pi} - F_{k-1}^{\pi} \tag{3.17}$$

Since probabilities of x_i in G_k^{π} are all the same, $U_k = G_k^{\pi}$.

where G refer to any subset within the nested structure.

Let x_i^k be an element in $U_k = G_k^{\pi}$ then $m_{\pi}(F_k^{\pi})$ is derived inversely when a probability distribution is given (Klir, 2006).

$$m_{\pi}(F_k^{\pi}) = |F_k^{\pi}| \ (p(x_i^k) - p(x_i^{k+1})) \tag{3.18}$$

where $p(x_i^{k+1}) = 0$

$$F_k^{\pi} = \bigcup_{h=1}^{k} U_h, \ k = 1, \dots, K \tag{3.19}$$

Equation (3.13) ensure the principle of order preservation

From the above, a possibility distribution is derived from the probability distribution in the following way:

$$\pi(x_i) = \begin{cases} \Pi(U_k), & \text{if } x_i \in U_k = F_k^{\pi} - F_{k-1}^{\pi} \\ 0, & \text{Otherwise} \end{cases} \tag{3.20}$$

$$\Pi(U_k) = \sum_{A \cap U_k \neq \varnothing} m_{\pi}(A) = \sum_{F_j^{\pi} \cap (F_k^{\pi} - F_k^{\pi}) \neq \varnothing} m_{\pi}(F_j^{\pi})$$

$$= \sum_{j=k, K} m_{\pi}(F_k^{\pi}) = \sum_{j=k, K} |F_j^{\pi}| \ ((p(x_i^j) - p(x_i^{j+1})))$$

$$= |F_k^{\pi}| \ p(x_i^k) + \sum_{j=k+1}^{K} |U_j| \ (p_{xi}^j) \tag{3.21}$$

Assume that elements in U are ordered so that $p(x_1) \geq \dots \geq p(x_n)$ and that

$U_k = \{x_a, \ldots, X_b\}$, $1 \le a \le i \le b \le n$. Then, the next Equation derived by Yamada (2001) shows.

$$\pi(x_i) = ip(x_i) + \sum_{j=i+1,n} p(x_j) \qquad (3.22)$$

As shown by Yamada (2001), the transformation satisfies both the consistency principle and the preference preservation principle.

Transformation of fuzzy variables

A fuzzy membership function where the alpha-cuts represent a set of nested intervals can be assumed a possibilistic body of evidence. If X is a fuzzy variable denoted with X_α and has generic alpha-cuts, $0 \le \alpha \le 1$, then alpha-cuts can be ordered in a nested structure. In this case, the necessity functions are special belief functions, and possibility functions are special plausibility functions. Hence, for the nested structure, it is possible to associate them with a basic probability assignment, a necessity function, and a possibility function in terms of alpha-cuts.

Figure 3.5 depicts the values of alpha limited to the finite set {0 0.1 0.2 ......0.9 1} and the following relationship applies:

$$X_{\alpha=1} \subset X_{\alpha=0.9} \subset X_{\alpha=0.80} \subset \ldots \subset X_{\alpha=0} \qquad (3.23)$$

In this case, any indices where $k = 1$ corresponds to $\alpha = 1$; $k = 2$ corresponds to $\alpha = 0.90;$....: and $k = n$ corresponds to $\alpha = 0$. This implies that alpha-cuts with $\alpha = 0$ includes all other alpha-cuts, where as the alpha-cut with $\alpha = 1$ is included by all others alpha-cuts, such that $0 \le \alpha \le 1$.

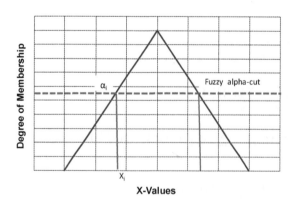

Figure 3.5 Generic alpha-cuts X_α and subset

For a triangular fuzzy membership function with a set of alpha-cuts and their negations, the following relations apply (as seen in Figure 3.5).

$$N(\overline{X}_{\alpha_j}) \equiv 0 \qquad (3.24)$$

$$\Pi(\overline{X}_{\alpha_j}) = 1 - N(X_{\alpha_j}) = \sum_{\alpha=0}^{\overline{\alpha}_j} m(X_\alpha) \qquad (3.25)$$

In Equation (3.18), $\overline{\alpha}_j$ denotes the value α, which proceeds the $\overline{\alpha}_j = \alpha_j - 0.1$ cut with $\alpha = 1$.

A fuzzy variable is represented by its membership function $\mu(x)$. It implies that $\mu(x)$ quantifies the value between 0 and 1. For the case of perfect evidence, as outlined in the previous section, this value also refers to the degree of belief, based on the available evidence, that element x will occur among all possible elements in X, or in other words, how much $\overline{x}$ is possible. Therefore, the membership function $\mu(x)$, for a given fuzzy variable X, exactly corresponds to the possibility distribution function $\pi(x)$, as defined by the possibility theory. Formally,

$$\pi(x) = \mu(x) \qquad (3.26)$$

for all, $\mu(x) \in X$

Using the above, because the smallest possibility distribution has the form $\pi(x) = (1, 0, ..., 0)$, the fuzzy variable that represents perfect evidence degenerates into a singleton. Similarly, because the largest possibility distribution has the form $\pi(x) = (1, 1, ..., 1)$, the fuzzy variable that represents total ignorance is a rectangular fuzzy variable. It follows that

$$\pi_i = \pi(x_i) = \alpha_i \qquad (3.27)$$

That is, the values assumed by the possibility distribution function at the extreme points of each alpha-cut is as shown in Figure 3.4. However, the possibility distribution function assumes different values in the different points of each alpha-cut. Still, it is always

$$max_{x \in X_\alpha} \pi_X(x) = 1, \quad \forall \alpha \qquad (3.28)$$

This can also be readily derived from a geometrical point of view. Considering Figure 3.4, the peak value of the reported fuzzy variable (i.e., unitary value of the membership function and the possibility distribution function too) is included in all alpha-cuts of the fuzzy variable. This implies the following relation,

$$\pi_i = \sum_{\alpha=0}^{\alpha i} m(X_\alpha)$$

(3.29)

It will equivalent to

$$\alpha_i = \sum_{\alpha=0}^{\alpha i} m(X_\alpha)$$

(3.30)

this more explicitly becomes,

$$\alpha = 0 = m(X_{\alpha=0})$$

$$\alpha = 1 = m(X_{\alpha=0}) + ... + m(X_{\alpha j}) + ... + m(X_{\alpha=1})$$

....

$$\alpha = 1 = m(X_{\alpha=0}) + ... + m(X_{\alpha j}) + ... + m(X_{\alpha=1})$$

It confirms that $x_1 = X_1, ..., x_n = X_n$ and $m(X_{\alpha=0}) = 0$ and m will be

$$m(X_{\alpha j}) = \alpha_j - \bar{\alpha}_j$$

(3.31)

From Equations (3.25), (3.30) and (3.31), two important relationships between the possibility and necessity functions and the values of levels of alpha follow:

$$\Pi(\bar{X}_{\alpha_j}) = \sum_{\alpha=0}^{\bar{\alpha}_j} m(X_\alpha) = \sum_{\alpha=0}^{\alpha_j} m(X_\alpha) - m(X_{\alpha_j}) = \alpha_j - (\alpha_j - \bar{\alpha}_j) = \bar{\alpha}_j$$

(3.32)

$$Nec(\bar{X}_{\alpha_j}) = 1 - Pos(\bar{X}_{\alpha_j}) = 1 - \bar{\alpha}_j$$

(3.33)

The basic probability assignment is equal for all alpha-cuts in cases where alpha-cuts are chosen equally spaced. As seen in Figure 3.5, the necessity function increases as the width of the alpha-cuts increases, that is, as alpha decreases. The possibility function of the negation of the alpha-cuts increases as alpha increases. This aligns with the fact that as the width of the alpha-cuts decreases, the total belief about the alpha-cuts decreases, and the possibility that something outside the alpha-cuts occurs becomes greater. The necessity function is always zero outside the alpha-cuts.

Based on the above relations established for the transformation of fuzzy sets into the probability and assumption that applies to DST, the following algorithms have been proposed:

Algorithms for probability to fuzzy transformation

i) Represent the uncertainty due to a random event with a probability distribution function

(PDF), either discrete $\sum_{i=1}^{n} p(x_i) = 1$ or continuous $\int_{\mathbb{R}} p_x \, dx = 1$.

ii) Considering a unimodal PDF, identify an interval $[X_L, X_R]$ in the PDF so that

$\int_{X_L}^{X_R} p(x) \, dx \approx 1$. The $[X_L, X_R]$ level will be selected such that it will be within the

99.7% confidence interval. For example, if $p(x)$ is a Gaussian function with standard

deviation, σ, the interval $[-3\sigma, +3\sigma]$ around the mean value will be chosen. The

probability function that decreases asymptotically to zero will be neglected for simplicity.

iii) Determine the peak level $[X_p, X_p]$ of the PDF such that $\int_{X_p}^{X_p} p(x) \, dx = 0$.

iv) The base interval $[X_L, X_R]$ of the PDF will be considered as the support of the

possibility distribution function $\pi(x)$ (which is a fuzzy number $\mu(x)$). It will represent

the α-cut at level $\alpha = 0$ of the membership function. The top level will correspond to the

α-cut, at level $\alpha = 1$.

v) Within the base interval $[X_L, X_R]$ and peak $[X_p, X_p]$ of the PDF, undertake n

numbers of finite divisions of nested intervals $[X_{Li}, X_{Ri}]$ and calculate the corresponding

probability values. Order the calculated values such that

$p(A_1) \geq .. \geq p(A_i) \geq ... \geq p(A_n)$.

vi) Calculate the possibility distribution (membership value) based on the evidence theory

such that $\pi(A_i) = i * p(A_i) + \sum_{j=i+1}^{n} p(A_j)$.

vii) Plot the fuzzy membership functions calculated from Equation in step-vi.

Figure 3.6 Algorithms for probability to fuzzy transformation

---

Algorithms for fuzzy to probability transformation

i) Represent the uncertainty associated with the fuzzy event using a membership function.

ii) Perform a fuzzy alpha-cut operation by selecting a value of α_i at random in $[0, 1]$ and determine the lower bound and upper bound for all α_i $(i = 1,....,n)$ and the cardinality of each alpha-cut. If the values of the alpha-cut are limited to the finite set $\{0, 0.1, 0.2,......,1\}$ and ordered in a nested structure, then the nested set will be

$$A_{\alpha = 1} \subset A_{\alpha = 0.9} \subset A_{\alpha = 0.80} \subset .... \subset A_{\alpha = 0} \equiv X$$

iii) Order the possibility distribution obtained from the alpha-cut such that

$$\pi(A_1) = 1 \geq .. \geq \pi(A_i) \geq .... \geq \pi(A_{x+1}) = 0 .$$

iv) Calculate the probability value based on the evidence theory such that

$$p(A_i) = m_p(\{A_i\}) = \sum_{k=k, K} m_x(F_k^x) / |F_k^x|$$

where $\forall A_i \in G_k^x = F_k^x - F_{k-1}^x$.

v) Plot the probability distribution functions calculated from the Equation shown in step-iv.

Figure 3.7 Algorithms for fuzzy to probability transformation

The application of the proposed algorithms are further illustrated and discussed in Birmingham case study application example.

3.3.2 Prorogation of uncertainty

A brief review on uncertainty propagation techniques has been presented earlier in Chapter 2. Discussion on uncertainty representation (Figure 3.1) and techniques for uncertainty analysis (Figure 3.2) was presented earlier in section 3.1. A simplified flow diagram for the uncertainty modelling for UWS is presented in Figure 3.8. As shown, the framework will adopt the hybrid approach where uncertain information are transformed into one form and propagated by probabilistic and fuzzy theories based techniques. MCS, LHS, Bootstrapping, and the fuzzy alpha-cut technique have been proposed for this research as they are well-tested and widely applied techniques in uncertainty analysis The following sections presents the algorithms developed for modelling a few of those techniques of uncertainty propagation based on this simplified flow diagram in the Mat Lab platform.

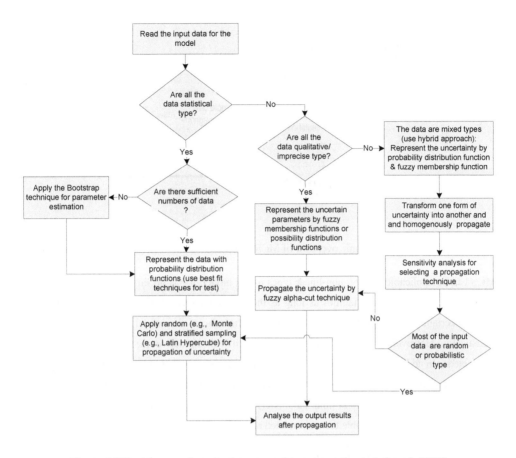

Figure 3.8 Decision tree for selecting uncertainty propagation technique in UWS

Monte Carlo simulation

A general introduction on Monte Carlo Simulation (MCS) is presented in Chapter 2. MCS is the most commonly used sampling method of uncertainty propagation for random types of input data. The MCS process assigns PDFs to each of the input parameters used for the analysis. Values for all parameters are then randomly selected from their PDFs, and a simulation is undertaken repeatedly. After a number of simulations, the output or result will form a distribution with an uncertainty. Figure 3.9 depicts the general process of using MCS in uncertainty analysis. The sample sizes are taken higher enough to ensure required accuracy in the output distribution functions (Van der Klis, 2003). Further analysis on sample size estimation is available in (Morgan and Henrion, 1990; Van der Klis, 2003).

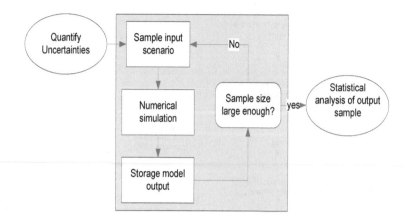

Figure 3.9 Schematic overview of MCS —modified from (Van der Klis, 2003)

MCS can be applicable for any non-linear types of the model. It has no limitations on the nature or magnitude of the input uncertainties as long as they can be described in statistical terms. If all the input uncertainties are described by PDFs, the information of the model output can be derived statistically. However, in many cases, it may be difficult to derive the distributions for input parameters. In addition, due to random sampling, MCS may have likely to be missing sampling data in some areas; therefore, it requires a large number of simulations. The large numbers of simulations for a complex model with several input parameters could have computationally difficult undertaking. Moreover, MCS is not applicable for uncorrelated random variables (Van der Klis, 2003).

Figure 3.10 illustrates the result of MCS applied for the head loss calculation in water systems for pipe diameter of 300 mm, pipe length 100 m, and relative roughness of pipes about 0.15 to 0.60. In this illustration, velocity is represented by a normal probability distribution function with a mean value of 1.40 m/s and a standard deviation of 0.20. The friction factor is represented by a uniform distribution function with a minimum value of 0.019 and a maximum value of 0.024. The results of 10,000 MCS shows that the average head loss in the system will be 0.75 m with the 95% confidence interval of 0.215 m and 1.49 m.

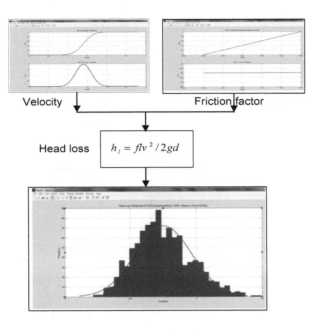

Figure 3.10 An example of Monte Carlo Simulation based uncertainty modelling

Latin hypercube sampling

Latin hypercube sampling (LHS) is a stratified sampling of the univariate distribution. This is accomplished by stratifying the cumulative distribution functions while equally and randomly sampling within the stratified areas. Many applications have demonstrated that LHS as a more efficient sampling method than MCS (McKay, 1992). Similar to MCS, LHS also requires PDFs to describe and propagate the uncertainties. Further discussion on LHS is available in Stein (1987); Helton and Davis (2003); Helton et al., (2006); Viana et al.,(2010).

A simplified procedure for LHS modelling is illustrated in Figure 3.11. This illustration shows two CDFs of two random variables with an equally divided sample size of 0.20. Here, n values obtained for X_1 (the first random variable) are paired in a random manner (of equally likely combination) with the n values of X_2 (the second random variable). These n pairs of $[X_1, X_2]$ are combined in a random manner with the n values of X_3 (the third random variable) to form the first N-triplets $[X_1, X_2, X_3]$ and so on, until the $(k-2)th$ N-triplets $(X_{1,}X_2, X_{n-2}, X_{n-1}, X_n)$ (k-number of random variables) are formed. The output probability distribution may then be approximated from the sample of K output values within each interval according to the probability distribution.

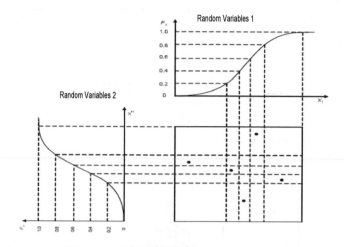

Figure 3.11 Schematic illustration of one possible LHS of size 5 for two variables

The example of head loss calculation presented earlier was also analysed using the LHS technique. It was processed for the same distribution functions with 0.10 size interval division. The result after 1000 simulations shows that the average head loss will be 0.74 m with 95% confidence interval of 0.15 and 1.58 m. Due to smaller numbers of inputs parameters and simple model, the calculation time was not different compared to MCS; however, the result shows a small variation.

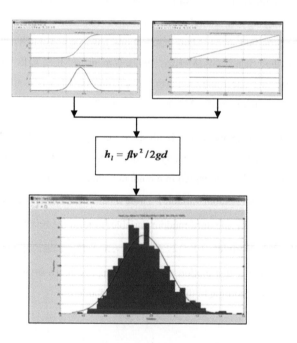

Figure 3.12 An example of LHS based uncertainty modelling

Fuzzy set theory based approach

If all the uncertainty parameters are imprecise or qualitative then either fuzzy membership functions or possibility distribution functions are used to describe the variables. The propagation of the uncertainty variables is achieved through the fuzzy alpha-cut technique. The fuzzy alpha-cut technique is a widely used approach in uncertainty analysis. Let fuzzy set $\tilde{A}$ with a triangular membership function (Figure 3.13), where the *alpha-cut* level intersects at two points a and b on the membership function. The values of variable x corresponding to points a & b are x_1 and x_2 respectively. As such, A_a contains all the possible values of the variables X between and including x_1 (the lower bound) and x_2 (the upper bound).

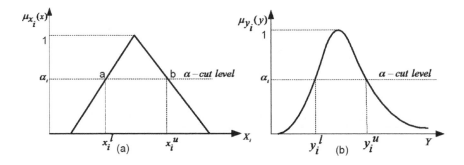

Figure 3.13 a) An example of a membership function of the input fuzzy variable X_i with its lower and upper bounds for a given *alpha-cut level*, and b) an example of a membership function of the output Y with the lower and upper bounds for a given *alpha-cut level*.

The fuzzy alpha-cut technique of uncertainty modelling was also applied for the head loss calculation example as presented earlier. The velocity and friction factor were represented by a triangular fuzzy membership function (Figure 3.14). The flow velocity is around 1.40 m/s (with a minimum value of 1.0 m/s and a maximum value of 1.80 m/s) with a friction factor of about 0.021 (with a minimum value of 0.019 and a maximum value of 0.024). The result reveals head loss to be around 0.70 with other values as [0.32, 0.70, 1.32]. The results of head loss calculations are not exactly equal to the MC and LHS but they are consistent.

Algorithms for fuzzy to probability transformation

i) Represent the uncertainty associated with the fuzzy event using a membership function.

ii) Perform a fuzzy alpha-cut operation by selecting a value of α_i at random in $[0, 1]$ and determine the lower bound and upper bound for all α_i $(i = 1,....,n)$ and the cardinality of each alpha-cut. If the values of the alpha-cut are limited to the finite set $\{0, 0.1, 0.2, ...,1\}$ and ordered in a nested structure, then the nested set will be

$$A_{\alpha=1} \subset A_{\alpha=0.9} \subset A_{\alpha=0.80} \subset .... \subset A_{\alpha=0} \equiv X$$

iii) Order the possibility distribution obtained from the alpha-cut such that

$$\pi(A_1) = 1 \geq .. \geq \pi(A_i) \geq .... \geq \pi(A_{n+1}) = 0 .$$

iv) Calculate the probability value based on the evidence theory such that

$$p(A_i) = m_p(\{A_i\}) = \sum_{k=k,K} m_\pi(F_k^\pi) / |F_k|$$

where $\forall A_i \in G_k^\pi = F_k^\pi - F_{k-1}^\pi$.

v) Plot the probability distribution functions calculated from the Equation shown in step-iv.

Figure 3.7 Algorithms for fuzzy to probability transformation

The application of the proposed algorithms are further illustrated and discussed in Birmingham case study application example.

3.3.2 Prorogation of uncertainty

A brief review on uncertainty propagation techniques has been presented earlier in Chapter 2. Discussion on uncertainty representation (Figure 3.1) and techniques for uncertainty analysis (Figure 3.2) was presented earlier in section 3.1. A simplified flow diagram for the uncertainty modelling for UWS is presented in Figure 3.8. As shown, the framework will adopt the hybrid approach where uncertain information are transformed into one form and propagated by probabilistic and fuzzy theories based techniques. MCS, LHS, Bootstrapping, and the fuzzy alpha-cut technique have been proposed for this research as they are well-tested and widely applied techniques in uncertainty analysis The following sections presents the algorithms developed for modelling a few of those techniques of uncertainty propagation based on this simplified flow diagram in the Mat Lab platform.

A simple algorithm proposed for bootstrap modelling to estimate the empirical cumulative distribution function of a statistic $\hat{\theta}$ from a given sample x_i, $i = 1, 2, \ldots, B$ $x_i \in X$ consists of the following steps:

i)   Store the available observed data in a database.

ii)   Resample each observed data point x_i with replacement with an equal probability of $1/n$

iii) The sample x_i continues to be resampled with replacement until B bootstrap samples $\hat{\theta}_i$

      $i = 1, 2, \ldots, B$ are obtained.

iv)   Each bootstrap sample x_i yields a bootstrap estimate of the statistic θ leading to the

      bootstrap estimates $\hat{\theta}_i$ $i = 1, 2, \ldots, B$, the desired result.

v)   The statistical information after B samplings will be used to identify the mean, standard

      deviation and other statistical characteristics of the observed data.

A modified to bootstrapping is moving–block bootstrap. The moving-block is typically used for seasonal data, such as such as daily temperature variation, monthly precipitation variation, daily water demand variation, etc. In such cases, a sampling technique should maintain the seasonal relation intact. The moving–block bootstrap has become well established since its development by Kunsch (1989) and Liu and Singh (1992).

In this technique, a single block of data is drawn at a time, instead of one x_i at a time, in order to preserve the underlying serial dependence structure that is present within the sample. In a moving-blocks bootstrap, a block length $L \square n/k$ is chosen where n is the record length and k is the number of blocks to resample. The idea is to choose a large enough block length L so that observations more than L time units apart will be nearly independent. For further discussion on the bootstrap technique and its application see Efron and Tibshirani (1993) and Burr (1994).

Application example

Table 3.1 presents the monthly average temperature simulated by the HadRM3 global climate model for Birmingham UK with a Special Report on Emissions Scenario (SRES)-A2 for the time slice 2020. The time slice 2020 provides a climate change prediction for 2011 to 2040. It includes four core emissions scenarios (Low, Medium-Low, Medium-High, and High) at 50 km resolutions. Now, the question is how to calculate the statistical information mean, standard deviation and confidence intervals for each month and season.

Table 3.1 Temperature forecasted by UKCIP02

Month	Prediction scenarios			
	High	Low	Med-High	Med-Low
Dec	0.81	0.68	0.76	0.76
Jan	0.71	0.6	0.67	0.67
Feb	0.69	0.58	0.65	0.65
Mar	0.76	0.64	0.71	0.71
Apr	0.88	0.74	0.82	0.82
May	1.03	0.86	0.96	0.96
Jun	1.22	1.03	1.15	1.15
July	1.45	1.22	1.36	1.36
Aug	1.57	1.32	1.47	1.47
Sep	1.48	1.24	1.38	1.38
Oct	1.23	1.03	1.15	1.15
Nov	0.97	0.82	0.91	0.91

In order to calculate the mean, standard deviation, and confidence interval of temperature in December from the predicted model results (a limited data case): 0.81, 0.68, 0.76 and 0.76, simple bootstrap modelling was used. After 1000 bootstrap resample, the mean, standard deviation, and a 95% standard deviation of temperature in December was 0.75, 0.0238, and 0.72 and 0.79 respectively (Figure 3.15).

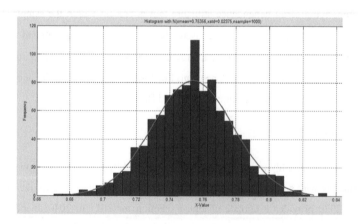

Figure 3.15 Bootstrap model results for statistical calculation

Figure 3.16 presents the average seasonal temperature in December-January-February using the moving-block bootstrap technique. The moving-block of the 1000 bootstrap resample generates an average value of 0.68 with a 95% confidence interval of 0.68 and 0.69.

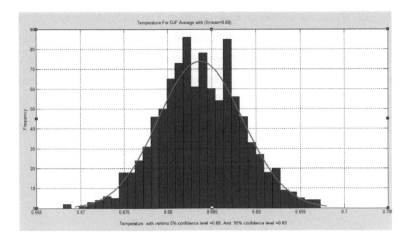

Figure 3.16 Moving block bootstrap model results for statistical calculation

3.5 Application of Hybrid Approach to Uncertainty Analysis

The main objective of this case study is to demonstrate the hybrid approach to uncertainty analysis while analysing the risk of future water availability in Birmingham, UK, for the year 2035. It analyses the potential future change pressures and associated uncertainties while forecasting the micro components of water demand. The results of this analysis are then compared with the results of water demand forecasted by Severn Trent Water Limited (STWL) for year 2035.

3.5.1 Introduction

Birmingham is the second most populated British city after London. It is situated at 52° 29′ 1″ N, 1° 54′ 23″ W, having a total area of 267.77 km2, an average elevation of 150 to 350 m, and a population of 1,113,600 (Birmingham City Council, 2008). The city of Birmingham forms part of the larger West Midlands conurbation and includes several neighbouring towns and cities, such as Solihull, Wolver Hampton, and the towns of the Black Country. Although Birmingham's industrial importance has declined, it has developed into a national commercial centre, being named the third best location in the UK to locate a for business, the 21st best in Europe, and the fourth most visited city by foreign visitors in the UK.

STWL and South Staffordshire Water (SSW) Company are responsible for supplying the drinking water, managing the wastewater, and environment in Birmingham. STWL provides water from the Birmingham Water Resource Zone (WRZ), one of six available resource zones (Figure 3.17). The Birmingham WRZ includes most of the city of Birmingham and is supplied by the Elan Valley reservoirs in mid Wales via a 70-mile long aqueduct. The deployable output of this zone is approximately 329.6 Mega litres per day (ML/d), and there is approximately 314.9 ML/d of water available for use. The estimated total population within this zone was 1,125,183 in 2007, and the service area has an average occupancy rate of 2.2 per household, with per capita water consumption at 146.98 litres (STWL, 2008).

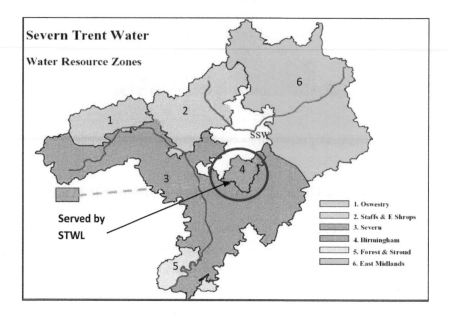

Figure 3.17 Severn Trent Water Resource Zones and supply areas for Birmingham

3.5.2 Analysis methods

Commonly applied water demand forecasting techniques are based on three approaches: end-use forecasting, econometric forecasting, and time series forecasting. The first approach predicts future water demand using data and assumptions based on current usage. The econometric approach statistically estimates the historical relationships between water consumption as a dependent variable and other factors as independent variables. This approach forecasts water demand by assuming that those relationships will continue into the future. In contrast, the time series approach forecasts water consumption in isolation, without forecasting the other factors on which water consumption depends. Using these approaches, Jain et al. (2001) developed linear regression models from weekly maximum air temperatures, weekly rainfall amounts, weekly past water demands, and the occurrence and non-occurrence of rainfall as dependent parameters. Bougadis et al.(2005) used these approaches to develop an Artificial Neural Network model as an improvement on regression technique. Altukaynak et al. (2005) proposed a Takagi Sugeno fuzzy method for predicting future monthly water consumption values from three antecedent consumption amounts. Billings & Agthe (1998) used the marginal price of water and personal income per capita in their model of future water forecasting. Zhou et al. (2000) model daily water consumption by splitting it into base (weather-insensitive) and seasonal (weather-sensitive) component uses for the Melbourne water supply. These approaches have also been used by a number of other studies (see, Arnell, 2004; Bradley, 2004; Arnell and Delaney, 2006).

In contrast to other studies, the focus of this study is explicitly on long-term water demand forecasting. As such, this water demand model necessarily includes a consideration of micro-demands of water, future population growth, climate change (temperature and precipitation), socio-economic changes, water loss by leakage, and the uncertainties associated with these factors (see Figure 3.18). It uses the hybrid approach to uncertainty analysis, where all the qualitative and quantitative information are presented and analysed in a single framework. It employs most of the algorithms developed and presented in the earlier sections that include conversion of fuzzy into the probability functions and vice versa, propagation by LHS, MCS, moving block bootstrapping and fuzzy alpha-cut technique. Due to limited water resources data, the case study computes the risk of water availability by comparing the analysed water demand with STWL water resources analysis. Based on the modelling framework (Figure 3.18), further analyses and results are presented in the following sections.

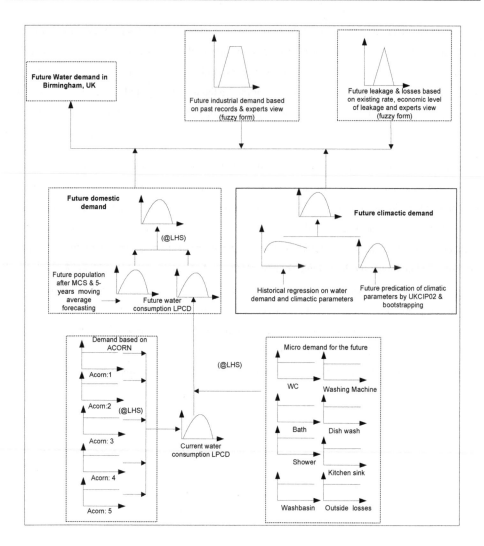

Figure 3.18 A flow diagram of analysis starting from the bottom upwards, where socio-economic analysis based on the classification of residential neighbourhoods (ACORN) on the bottom left, and micro-components of household water demand on the bottom right are used to forecast the future domestic water demand, and combined with the other demand including climatic factors, industrial demand and water losses.

3.5.3 Data availability and analysis

Historical total water demand

The time series record of total water consumption in Birmingham, served by STWL from January 1996 to March 2005 is shown in Figure 3.19. The record shows that water demand decreased from

2111 ML/month in January 1996 to 1837 ML/Month in January 1998. According to STWL, a potential reason for this decrease could be a reduction in the number of industries since 1996. There is an average rate of 1943 ML/month from 1998 to 2001 with no considerable increase in water demand. The average water consumption during the winter appears consistent, though summer time consumption figures are well above normal. Apparently, the peak demand increased in the warmest summer month of the year compared to normal years such as 1996. It is noted that the water consumption data for residential, commercial, or industrial purposes were not available. For ease of analysis, we have aggregated the data into monthly totals and normalized each total so that all data assume 30.416 days in a month.

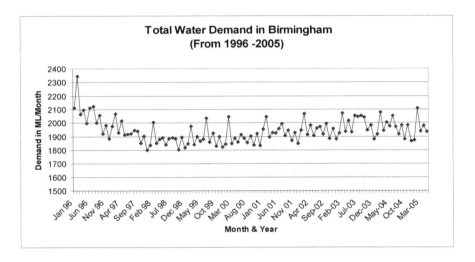

Figure 3.19   Total water consumption per month in Birmingham served by STWL

Historical and future climate change scenarios

Water demand is directly affected by rainfall patterns, air temperature, sunshine duration, relative humidity, and wind speed. However, it is very difficult to collect such data as required and to correlate that data with water demand. Due to these limitations, only temperature and precipitation are considered in the analysis. Historical time series records of the temperature and precipitation parameters for two rain gauge stations (Sutton Bonnington, Nottinghamshire and Penkridge, Shropshire) was provided by STWL. The average values observed on a monthly basis in these stations from January 1996 to May 2005 were used for this case.

In order to correlate the effects of temperature and precipitation on water demand, a linear regression model was developed using the data from January 2001 to March 2005, as shown in Figure 3.20 (a) & (b). According to the regression results, water demand is negatively correlated with precipitation

$Y=1.9934*10^6-0.0006*P$ with an adjusted regression coefficient of $R^2 = 0.46$ and standard error of 0.3369. Temperature is positively correlated $Y=1.9389*10^6+0.0021*T$ with an adjusted regression coefficient of $R^2 = 0.49$ and standard error of 0.0533. The regression results show a lower correlation of variables to the prediction. However, the correlation results indicate a rough relation of the temperature and precipitation with the total water demand.

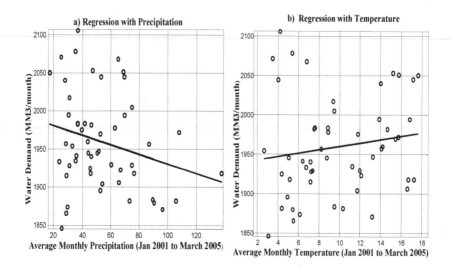

Figure 3.20 Regression results on water demand and (a) average precipitation and (b) temperature

Future climate change scenarios for Birmingham are based on the scenarios developed and studied by the UK Climate Impacts Programme-2002 (UKCIP02). UKCIP02 uses the Hadley Global Climate Model (HGCM) of the atmosphere, which includes a dynamic ocean circulation model (see www.ukcip.org.uk). It provides a climate change database for three time slices: 2020s (predictions for 2011 to 2040); 2050s (predictions for 2041 to 2070); and 2080s (predictions for 2071 to 2100). It also includes four core emissions scenarios (Low, Medium-Low, Medium-High, and High) at both 50km and 5km resolutions. For this case study, we use the temperature and precipitation parameters for four emissions scenarios for the 2020 time slice at 5 km resolutions (see Tables 3.2 and 3.3).

The climatic component of the future water demand is forecasted by considering the future precipitation and temperature changes as predicted by the UKCIP02 (Tables 3.2 and 3.3). The seasonal average of temperature and precipitation is analysed by using the moving block bootstrap method as discussed earlier. For the moving block bootstrap, the data were grouped into three months lengths for each of the four scenarios. After 1000 runs of moving block of bootstrapping, the seasonal average values with 95% confidence interval are presented in Table 3.4.

Table 3.2 Monthly average temperature and changes in mean temperature simulated by HadRM3 with Special Report on Emissions Scenarios (SRES) A2 for time slice 2020

Scenarios/Month	Dec	Jan	Feb	March	April	May	Jun	July	Aug	Sep	Oct	Nov
Base Average Temp °C	4.55	4.85	5.89	8.21	11.41	14.34	16	15.8	13.4	9.87	6.75	4.98
Low	0.69	0.63	0.6	0.63	0.69	0.77	0.89	1.05	1.15	1.1	0.95	0.8
Medium to Low	0.77	0.7	0.67	0.71	0.77	0.86	1	1.17	1.28	1.23	1.06	0.89
Medium to High	0.77	0.7	0.67	0.71	0.77	0.86	1	1.17	1.28	1.23	1.06	0.89
High	0.82	0.75	0.72	0.75	0.82	0.92	1.06	1.25	1.36	1.31	1.14	0.95

Table 3.3 Monthly average precipitation rate (mm/month) and changes in total precipitation rate (%) simulated by HadRM3 with SRES A2 for time slice 2020

Scenarios/Month	Dec	Jan	Feb	March	April	May	Jun	July	Aug	Sep	Oct	Nov
Base case Average	48.12	44.32	45.72	55.98	53.08	45.67	40	40.73	47.5	52.79	52.29	48.17
Low	4.93	5.41	3.68	0.91	-2.25	-5.41	-8.34	-10.4	-9.98	-6.64	-2.16	2.02
Medium to Low	4.09	1.01	-2.5	-6.02	-9.29	-11.6	-11.1	-7.4	-2.4	2.25	5.49	6.03
Medium to High	4.09	1.01	-2.5	-6.02	-9.29	-11.6	-11.1	-7.4	-2.4	2.25	5.49	6.03
High	5.87	6.44	4.37	1.08	-2.67	-6.43	-9.92	-12.4	-11.9	-7.9	-2.57	2.4

Table 3.4 Seasonal average changes in temperature and total precipitation after bootstrap sampling

Parameters/Month	Dec, Jan, Feb		March, April, May		Jun, July, August		Sep, Oct, Nov	
Lower 5%								
Tempt C	0.6		0.63		0.89		0.8	
Precipitation %	-2.5		-11.59		-12.38		-7.9	
Mean Value								
Tempt C	0.71		0.77		1.14		1.05	
Precipitation %	2.29		-6.86		-8.1		2.26	
Upper 95%								
Tempt C	0.82		0.92		1.36		1.31	
Precipitation %	6.44		0.91		-2.4		6.03	

Population growth

The population growth in any city depends on many factors including fertility rate, mortality rate, migration and other economic factors. Analysis of all those factors in detail is beyond the scope of this study. This study uses the future trends of total population forecasted by the Birmingham City Council for the years 2006 to 2035. STWL supplies to a partial area of Birmingham city and includes additional other neighbouring towns beyond Birmingham city. However, the trends of the total population are assumed to be the same.

The probabilistic approach was employed to represent the uncertainty associated with population growth and population forecasting. The demographic data of total population obtained from the city council is used to find the best distribution of total population growth. From the statistical analysis of existing population growth, a lognormal distribution is identified as the best fit to represent Birmingham's total population. Therefore, we apply a lognormal distribution to forecast the population in the STWL service area. The future population for the year 2035 is then forecasted using 1000 runs of Monte Carlo simulation with 5-year time steps smoothened moving average method (see, Brockwell and Davis, 2002). The validated historical population trends with 95% confidence interval for the year 2035.using moving average methods and MCS are shown in Figure 3.21 (a) and Figure 3.21 (b)

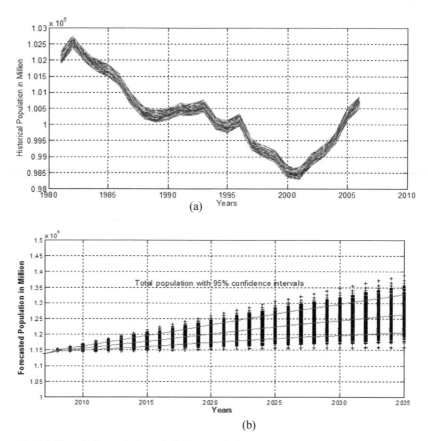

Figure 3.21 (a) Historical population of Birmingham validated by MCS and moving average method
(b) Total population forecasted for Birmingham served by STWL

Socio-economic change

The impact of socio-economic pressures on future water demand in Birmingham is analysed using geo-demographic classification. A classification of residential neighbourhoods (ACORN) is a geo-demographic segmentation of the UK population. It groups small neighbourhoods, postcodes, and consumer households into 5 Categories, 17 Groups, and 56 Types (see www.caci.co.uk). ACORN is used to understand consumer lifestyles, behaviours, and attitudes, the needs of the neighbourhoods, and the people's public service needs.

STWL observed the water consumption patterns for households categorized from ACORN 1 to ACORN 5 for family sizes of 1 to 5 for **888** families in Birmingham. According to their household classification, ACORN 1 refers to *"wealthy achievers"*; ACORN 2 refers to *"urban prosperity"*; ACORN 3 refers to *"comfortably off"*; ACORN 4 refers to *"moderate means"*; and ACORN 5 refers to *"hard pressed"*. The lower ACORN numbers are associated with wealthier and more educated people. There are other sub-classifications within each ACORN, but these are not considered in this analysis (see www.caci.co.uk for more information). Figure 3.22 presents the water consumption patterns per capita per day for the different ACORN types and their respective occupancy sizes. As expected, average water consumption rate is varying with ACORN and occupant size.

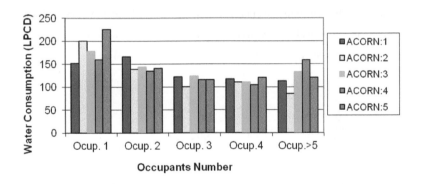

Figure 3.22 Per capita water consumption based on ACORN

Deterioration of infrastructure systems

The average age of water mains in Birmingham is 60 years (STWL, 2008). Because of the ageing and deteriorating of water mains, pipes failure rates are high and resulting water leakage and losses. It has been reported that over 25% of the city's treated water is lost through leaking cracks and breaks in the mains. In 2008, gross loss by pipe leakage was reported to be 90 Ml/day (STWL, 2008).

Identifying economic leakage rates for the future is a complex undertaking. It depends on many factors, including water price, the costs of investments for pipe rehabilitation, and existing regulations that must be met—for example, regulations set by the office of water services or the Environmental Agency in the UK. Furthermore, increasing future water demands combined with pipe ageing will result in increased leakage rates in the future. For this case, the targets for economic leakage in Birmingham have been calculated using least-cost modelling, which aims for linear reduction in leakage rates over a 5 year Annual Maintenance Plan (AMP5). The projected future economic levels of leakage in Birmingham based on the STWL source is presented in Table 3.5, below.

Table 3.5 Economic level of leakage for Birmingham based on STL, Birmingham

Area	Units	AMP4	AMP5					AMP6	AMP7	AMP8	AMP9
		2011	2011	2012	2013	2014	2015	2020	2025	2030	2035
Birmingham	Ml/d	76.0	76.0	76.0	76.0	75.9	75.9	71.8	69.4	72.8	71.5

STWL used this targeted leakage quantity for their future water demand prediction. However, it is clear that total future water leakage from the systems is an uncertain quantity that depends on many factors including the rate of rehabilitation of pipes, environmental condition, and flow characteristics of the pipes. In this case, this is represented by a triangular shaped fuzzy membership function with the most likely value of 80 ML/d, minimum 70 ML/d, and maximum of 90 ML/d respectively as shown in Figure 3.23. It is noted that the ranges and likely values of water losses was decided after the consultation and discussion with the experts working at STWL, Birmingham office. The most likely value is the targeted water losses rate from the systems planned by STWL.

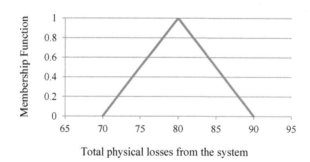

Figure 3.23 Total physical losses of water from the system (Ml/d) in a fuzzy form

Forecasting the domestic and industrial water demand

The domestic water demand was forecasted considering socio-economic condition (i.e., ACORN) and micro-demand set by the Environmental Agency, UK. The water demand per capita per day (LPCD) for each of the ACORN is presented in Figure 3.22. According to this ACORN record, the average and standard deviation of water consumption rate for each ACORN are ACORN 1: (182.9, 30.2); ACORN 2: (145.01, 10.50); ACORN 3: (115.10, 12.20); ACORN 4: (112.0, 11.20): ACORN 5: (121.0, 23.50). Using these figures, a normal distribution functions for 5 categories of water demand were developed. The daily water consumption was then calculated by Latin hypercube sampling technique. The resulting average water demand after 1000 simulations is 146.50 LPCD with standard deviation of 10.98 and 95% confidence interval of 127.79 LPCD to 164.45 LPCD as shown in Figure 3.24 (a).

The Environmental agency in the UK has classified the micro-components of daily water consumption into 8 groups which are: Water Closet: (10-40%); Washing Machine: (10-20%); Dish Washing: (5-8%); Bath: (10-20%); Shower: (4-8%); Kitchen sink: (8-20%); Washbasin: (4-12%); Outside use and losses: (10-15%). In this analysis, the ranges of eight micro-components of daily water consumption were represented by the uniform PDFs. The future water consumption per capita per day has been forecasted using the ACORN based water consumption rate and sampling the PDFs of micro-demand by using the Latin hypercube sampling. The result after 1000 simulation shows the future water demand of 141.37 LPCD with standard deviation of 16.71 and 95% confidence interval range 114.36 to 169.77 LPCD (see Figure 3.24 (b)).

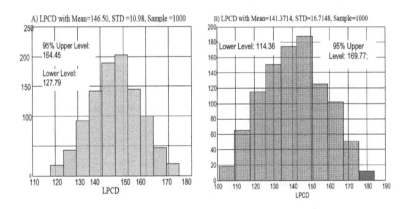

Figure 3.24 Water consumption based on ACORN (a) for the present and (b) for year 2035

As discussed earlier, STWL does not retain any historical records on industrial and other sectors demand. In this case, we consider future industrial demand as an uncertain quantity and has been represented by a trapezoidal shaped fuzzy membership function as shown in Figure 3.25. The potential future demand is given by [45, 50, 60, 65] estimated after consultation with the experts working at STWL.

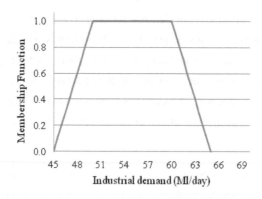

Figure 3.25 Representation of total daily industrial water consumption in Birmingham (ML/d) by a fuzzy membership function

3.5.4 Hybrid approach to uncertainty analysis

The future water demand is analysed based on the framework shown in Figure 3.18. The domestic water demand is forecasted considering the per capita water demand (Figure 3.24) and total population (Figure 3.21), which is represented in a probabilistic form. The leakage and losses from the water supply system is represented in a fuzzy form (Figure 3.23). The impacts of climatic factors on total demand are considered using the regression results (Figure 3.20). The regression models roughly show that there will be a 2.10% increase and 0.06% reduction in total water demand, if temperature increases by 1°C, and if rainfall increases by 1mm/month respectively. If we consider only increase in the temperature, expected increase of average water demand will be by 7.52 ML/day with standard deviation of 2.51.

The total water demand in Birmingham for the year 2035 was analysed applying the algorithms developed from this study. The algorithms have been applied in the following cases:

I)   Transformation of fuzzy information into the PDFs and vice verse;

II)   Future water demand analysis by

    a)   Transformation of all the input parameters into PDFs and propagating the future water demand by Latin hypercube sampling technique;

    b)   Transformation of all the input parameters into FMFs and propagating the future water demand by Fuzzy alpha-cut technique;

III) Comparison of the results with other hybrid and homogenous methods of uncertainty analysis;

I)   Transformation of all the fuzzy information into the PDFs and vice verse by applying the developed algorithms for the transformation

The main variables considered for the analysis are domestic water demand, industrial water demand, and water leakage and losses from the systems. As shown earlier, the domestic water demand has been calculated considering the micro-components, which is in a probabilistic form. Other parameters, such as climatic demand is also represented in probabilistic form; whereas other two components are represented in a fuzzy form (Figure 3.26). In order to demonstrate the conversion process, the algorithms presented in earlier sections in Figure 3.6 and Figure 3.7 are applied.

To convert a fuzzy membership function to the PDF will require the application of algorithm shown in Figure 3.7. For example, water leakage in Birmingham is shown with membership values of [70, 80, 90]. Using this membership value and taking alpha-cut of 0.10 (that will result 10 alpha-cuts) and application of algorithm of Figure 3.7 will transform the fuzzy membership function into the PDF. The transformed PDFs from the fuzzy membership functions are shown in Figures 3.26 (a) and (b). The transformed PDFs have maximum value where the possibility value of fuzzy membership function is very likely. Similarly, the order of transformation has been maintained in the transformed distributions function.

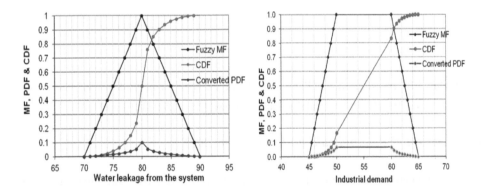

Figure 3.26 Transformation of MFS into PDFs: (a) total daily physical leakage and losses (ML/d) and (b) industrial water demand (ML/d)

The histogram for the domestic water demand is shown in Figure 3.27. The PDF of the resulting histogram and conversion into the fuzzy form requires application of algorithm given in Figure 3.6. When applying this algorithm the characteristic of a PDF is required , such as the most likely value (i.e. the mode of the distribution function) and 99.7% confidence interval for the PDF are mean: 178.76, STD: 20.86, lower: 145.00 and upper: 212.54 respectively. After the application the converted results of PDFs into MFs are shown in 3.28 (a) and 3.28 (b).

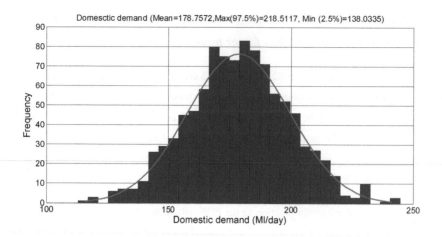

Figure 3.27 Total daily domestic water demand in Birmingham (ML/d) with 97.5% confidence
interval

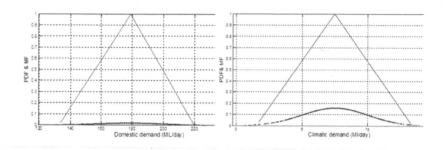

Figure 3.28 Transformation of PDFs into MFS (a) total daily domestic demand [138.03, 178.75
218.51] (ML/d) and (b) climatic demand [2.51, 7.52, 11.93] in (ML/d)

II) *Total water demand calculation for year 2035*

(a) Transformation of all the input parameters into the PDFs and propagating the future demand by Latin hypercube sampling technique

In this case, water losses, and leakage from the systems and industrial water demand are
transformed into the PDFs (Figures 3.26(a) & (b)). All the PDFs are then propagated using the
LHS algorithm and with 1000 simulations. The results of the analysis are shown in Figure 3.29.
As shown, the average demand for year 2035 is 323.05 ML/day, with STD of 22.45 and 95%
confidence interval of 285.00 ML/day and 359.75 ML/day.

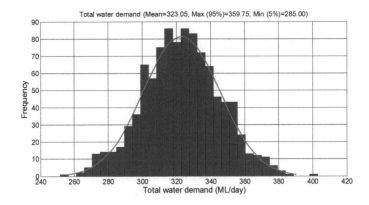

Figure 3.29 Total daily domestic demand calculated by transforming all the FMS into the PDFs and LHS

(b) Transformation of all the input parameters into FMFs and propagation of future demand by fuzzy alpha-cut technique

The future domestic water demand and climatic demand are converted into the fuzzy membership functions. All the FMFs are then propagated homogenously using the fuzzy alpha-cut technique with 100 samples. The result of the analysis is shown in Figure 3.30. The result shows the future water demand in a trapezoidal shaped membership function [254.4, 317.10, 326.20, 386.70] Ml/day. As shown, the most likely future water demand is 321.65 ML/day.

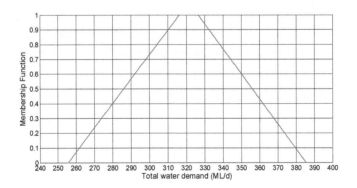

Figure 3.30 Total daily domestic demand calculated by transforming all the PDFs into the FMFs and propagating by fuzzy alpha-cut technique

95

III) Comparison of the results with other hybrid and homogenous methods of uncertainty analysis

The analysed results are compared with: i) hybrid approach proposed by Guyonnet et al. (2003) and ii) conventional homogenous approach where all the input parameters are represented by PDFs and propagated by MCS. Both the techniques are very common in risk analysis and decision making problems.

(a)   Application of the hybrid algorithms developed by Guyonnet et al. (2003)

The hybrid method proposed by Guyonnet et al. (2003) combines the random probabilistic variables with possibilistic (here fuzzy) variables. It consists of repeated Monte Carlo sampling to process the uncertainty associated with the probabilistic variables and fuzzy interval analysis (fuzzy alpha-cuts) to process the uncertainty due to possibilistic variables. If a model consists of multiple inputs parameters with probabilistic variables, $(P_1, P_2, ..., P_n)$ and possibilistic variables, $(F_1, F_2, ...F_m)$ then the model output, $f(.) = f(P_1, P_2, ..., P_n, F_1, F_2, ...F_m)$ is analysed using the following steps (Guyonnet et al., 2003):

i)   Generate i^{th} realization of n probabilistic variables vectors: $P_1, P_2, ..., P_n$. In this case, the domestic demand and climatic demand are considered as the probabilistic variables.

ii) Select a value alpha-cut of the membership functions and corresponding cuts of the membership functions. The leakage and losses from the systems and industrial demand are considered as the fuzzy variables.

iii) Calculate the inf (smallest) and sup (largest) values of the alpha-cut.

iv) Assign these inf and sup values to the lower and upper limits of the alpha-cuts of $f(.) = f(P_1, P_2, ..., P_n, F_1, F_2, ...F_m)$.

v)   Return to step (ii) and repeat steps (iii) and (iv) for another alpha-cut. The fuzzy result of $f(.) = f(P_1, P_2, ..., P_n, F_1, F_2, ...F_m)$ is obtained from the inf and sup values of $f(.) = f(P_1, P_2, ..., P_n, F_1, F_2, ...F_m)$.

vi) Repeat to step (i) to generate a new realization of the random variables.

After 'N' times simulation, a family of 'N' fuzzy interval values is thus obtained (Figure 3.31). The final inf and sup values of $f(.) = f(P_1, P_2, ..., P_n, F_1, F_2, ...F_m)$ is developed after extracting a certain percentage of confidence level for each level of alpha-cut. For each alpha-cut of random fuzzy set, all the left side and right side of sets are arranged in an increasing order. The set $[Finf_d^{\alpha}, F\sup_d^{\alpha}]$ is considered such that $P(leftside \leq F\inf_d^{\alpha}) = 1 - d\%$ and $P(rightside \leq F\sup_d^{\alpha}) = d\%$. A fuzzy interval F_d is formed within $\alpha \in [0,1]$. The standard value for d=95 is chosen to ensure the 95% confidence interval. The resulting total water demand for the future is shown in Figures 3.31.

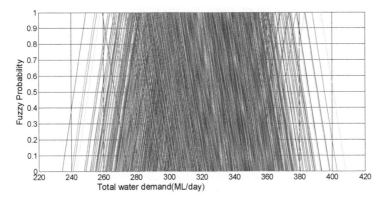

Figure 3.31 Total water demand forecasted using the hybrid approach suggested by Guyonnet et al., 2003

The resulting output is trapezoidal shaped with the future water demand of [266.74, 282.20 360.30 376.20] Ml/day (Figure 3.32). As shown, the most likely future water demand will be 321 ML/day.

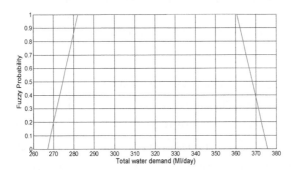

Figure 3.32 Total water demand forecasted using the hybrid approach suggested by Guyonnet et al.

(b) Total water demand in year 2035 in Birmingham - when analysed with the probabilistic method

The conventional probabilistic approach to uncertainty analysis is to represent all the input parameters by the PDFs and then propagated by LHS or any other simulation techniques. In this case, a uniform distribution function is assigned for the parameters, which has limited information. In this analysis, we assign uniform distribution functions for Industrial demand and water leakage from the systems. The minimum and maximum corresponding values for industrial water demand will be [40, 60]

ML/day and water leakage and losses from the systems will be [70, 80] ML/day. The result of the analysis is shown in Figure 3.33.

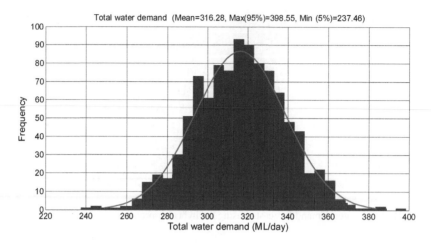

Figure 3.33 Total daily domestic demand considering all the distribution as PDFs

The comparative results of the future water demand analysis by different approach of uncertainty propagation and deterministic approach analysed by STWL is presented in Table 3.8.

Table 3.1 Comparative results of future water demand analysis in Birmingham for 2035

SN	Methods	Results
1	Hybrid by conversion of fuzzy membership functions into the probability distributions functions and homogenous propagation by LHS (Figure 3.29)	Mean: 323.04 ML/day, STD: 22.04, 95% confidence interval [285.00 359.75] ML/day
2	Hybrid by conversion of probability into the fuzzy membership functions and homogenous propagation in PDFS by fuzzy alpha-cut technique (Figure 3.30)	Most likely value: 321.65 ML/day, Value ranges: [254.4, 317.10, 326.20, 386.70] ML/day.
3	Hybrid approach proposed by Guyonnet et al., 2003 (Figures 3.31 and 3.32)	Most likely value:321 ML/day Value ranges: [266.74, 282.20, 360.30 376.20] ML/day.
4	Homogenous and probabilistic analysis (Figure 3.33)	Mean: 316.28 ML/day, STD: 22.47, 95% confidence interval [278.75, 353.39] ML/day.
5	Prediction by STWL (deterministic approach)	296.54 ML/day

- Forecasted mean value of total water demand from each of the techniques are similar, but all of them are higher than the STWL analysis. This could be because (i) the STWL analysis did not consider the climatic impact; (ii) they assumed that the industrial demand and losses from the systems to be almost constant for each year. The deterministic analysis of STWL, which cannot indicate the confidence interval of their future water demand, will not be sufficient to analyse future strategies and risk management.

- The envelop of uncertainty by hybrid approach with fuzzy approach and hybrid approach proposed by Guyonnet et al., 2003 are wider compared with the probabilistic approach. The likelihood level of Guyonnet et al. is wider than homogenous approach.

- The time of processing in hybrid approach with homogenous fuzzy operation is fastest among all.

3.5.5 Risk calculation

The risk of water availability in the future is estimated by comparing the forecasted water demand with the available existing water resources in Birmingham. As discussed, the water demand considered for the analysis include domestic demand, industrial demand, leakage and losses, and climatic demand. Figure 3.34 presents the total water demand in Birmingham for year 2035 with uncertainty. The envelope of uncertainty increases as the length of the prediction increases.

The current water resource capacity of the STWL Birmingham zone is 314. 90 ML/day. Based on the forecasted results, the existing resource will be able to meet the future water until year 2022 - if water demand increases along the average rate.

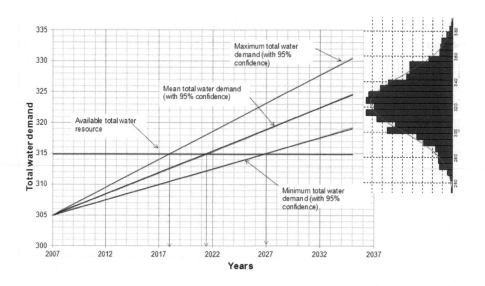

Figure 3.34 Total daily total water demand in Birmingham for the year 2035 with 95% confidence interval

The risk of water availability in Birmingham is processed by applying the Equations and other framework presented earlier in this Chapter. Basic risk Equation implies that risk is the cumulative impact of water scarcity to human life and environment. However, this analysis considers only population influenced by water scarcity in year 2035.

As shown in Figure 3.3.4, by 2035, the daily water demand in Birmingham will not be met by the existing sources. The mean, and maximum values of the uncertainty envelope show that Birmingham will have the water shortage of 2.58% (i.e., 8.15 ML/day), and 14.24% (i.e., 44.85 ML/day), respectively. The likelihood of risk is graphically plotted in the likelihood membership functions as shown in Figure 3.35. As shown, the likelihood values [0, 02.58/100, 14.24/100] is mapped into the 7-tuple fuzzy set using the maximum operation technique (Klir and Yuan, 1995). The resulting fuzzy number after the operation is [0.87, 0.39, 0, 0, 0, 0, 0]. Thus, the resulting likelihood of the risk of failure, after the normalisation, is [0.70, 0.30, 0, 0, 0, 0, 0] and shown as a dotted line in Figure 3.35. In several modelling cases, the likelihood of the risk could be presented in the probabilistic form. In such cases, PDFs results have to be transformed into FMFs applying the transformation technique developed and presented earlier.

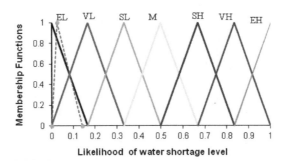

Figure 3.35 Likelihood of risk for water availability from the existing source

The severity component of the risk Equation represents the number of people affected by system failure. As presented earlier, the future water demand for year 2035 will be [285.00, 323.05, 359.75] ML/day. If we consider the existing water sources (314.9 ML/day), the average water deficit in year 2035 will be 8.15 ML/day. Given the future daily water consumption rate of 114.36 LPCD, the deficit will affect approximately 71,275 people daily. These values are normalized with the dimensionless interval of [0, 1] such that the normalized severity level can be represented by the fuzzy numbers defined in [0, 1]. Thus, the severity values with respect to the affected percentage of the population to the total population within 95% confidence interval [1179790, 1218173, 1260246] will result 6.041%, 5.85% and 5.65% respectively. This will result the normalised severity values of [0.06, 0.058, 0.056] respectively as presented and shown by dotted red lines in Figure 3.36.

According to the severity level classified in this research, the affected population will be in a range of the extremely low and very low. Using the same operation as for the likelihood level analysis, the normalised fuzzy values of the triangular fuzzy membership function will be [0.60, 0.40, 0, 0, 0, 0, 0]. Often, the severity of the risk will be subjective and may require expert support. In such cases, the average values from multiple experts can be calculated by the fuzzy operations as presented in Appendix III.

The magnitude of risk of water availability in Birmingham in 2035 is quantified using the risk Equations (see appendix III). The magnitude of the risk is estimated by fuzzy alpha-cut. Both likelihood of risk (Figure 3.35) and severity (Figure 3.36) are very low and thus, without further analysis, the resulting risk will be very low. The resulting magnitude is mapped to the risk classification and magnitude of the risk as obtained by the fuzzy max-operation.

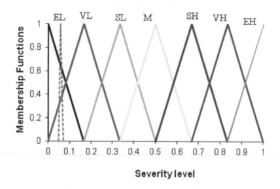

Figure 3.36 Level of severity from the unmet future demand) from the existing water source

Based on the risk classification, the potential risk of water availability in Birmingham lies in a very low level. There is not well-accepted measure of risk level in the literature. Our classification was general for the UWS and in this case, risk seen in a very low seems quite conservative. Often, the local community and utility managers define level of risk to be considered. It is clear that Birmingham must ensure 100% coverage for the entire population. It is very likely (with 95% confidence) that the existing water resource in Birmingham will not meet future demand even if the current usage patterns are maintained. Therefore, the utility manager must consider alternative options for augmenting the existing water source through risk management. Detailed analysis of alternative sources of water available to Birmingham is beyond the scope of the study.

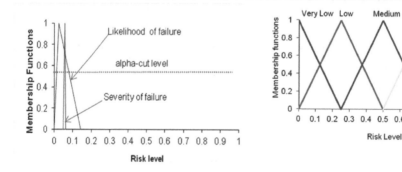

Figure 3.37 Risk calculation (shows both likelihood and severity levels for fuzzy alpha-cut technique) and risk classification

3. 6 Discussion and Conclusions

This Chapter presented a review of hybrid approach to uncertainty analysis; developed a new hybrid approach to uncertainty analysis; proposed several algorithms for the uncertainty modelling; developed fuzzy set theory based risk modelling, and the demonstration of the developed methodologies for risk of water availability in Birmingham UK in year 2035.

The literature review showed that the hybrid approach to uncertainty modelling allows us to analyse both aleatory and epistemic uncertainty in a single framework. This could be achieved either by (i) transforming all the information to probabilistic or fuzzy form or propagated homogenously through a model (i.e., either by a random simulation or fuzzy approach), or (ii) describing all the information separately and propagated heterogeneously.

Either of two options is possible in hybrid modelling. The first type requires the transformation of one form of uncertainty information to the other. These methods could have potential losses or gain of the information during the transformation operation. Therefore, there are many established principles to be followed during the transformation process to ensure that information is neither lost nor gained. Some of those principles discussed in the literature include Zadeh consistency principle (Zadeh, 1978), Dubois and Prade's consistency and preference preservation principles (Dubois and Prade, 1982), Delgado and Moral's maximal specificity principle (Delgado and Moral, 1987), Geer and Klir's information-preservation principle (Geer and Klir, 1992). Therefore, this approach requires special techniques to convert the information faithfully and apply the other commonly used methods of propagation.

The second approach to uncertainty analysis is often named referred to as a *hybrid approach* in literature. This approach attempts to describe and propagate the uncertainty heterogeneously and process the output with special treatment. A few of those commonly used techniques are (i) based on the concept of hybrid number by Kaufmann and Gupta (1985); Ferson and Ginzburg (1996); Cooper et al. (1996); or (ii) a joint hybrid propagation technique developed by Guyonnet et al. (2003), Baudrit and Dubois (2005) , Baraldi and Zio (2008). In joint hybrid propagation technique, the randomness is propagated separately using MCS and impreciseness by the fuzzy interval analysis. The resulting fuzzy random variable is further analysed with special procedures. This method in particular, output analysis was further improved by other researcher, such as (Baudrit et al., 2006; Baudrit et al., 2007).

Those techniques are very rich in scientific content and are well discussed and cited in the literature. The first types are requires additional process for transformation but propagation and output analysis is relatively easy. The second type is relatively easy to model; however; they require a separate mathematical operations for the description and propagation and treatment for the output. The result of this type of analysis is found to be conservative. As a result, the hybrid technique of heterogeneous propagation has a limited application in real cases modelling.

Considering the complexities of UWS, we propose the first approach of hybrid analysis. In this approach, all the information is transformed to one form (i.e., either in probability or fuzzy form) by applying the transferring algorithms that is based on Dempster-Shafer theory. The proposed transformation is applicable for continuous functions, fuzzy membership functions and ensures most of the principles of the transformation. The transformed uncertain information are propagated by MCS, LHS, and fuzzy alpha-cut technique. It is also noted that the proposed approach is very simple, and does not require intensive computing efforts to achieve a similar level of result analysis as with compared with other types of hybrid techniques.

The developed algorithms and models were applied for uncertainty analysis while modelling future water demand in Birmingham for year 2035. The same analysis was done by applying the algorithms developed by Guyonnet et al. (2003) and the probabilistic approach assuming all the probabilistic information without transformation. The comparative results show that: (i) the results of our approach is very consistent compared to other approach, (ii) the fuzzy approach requires the least computing time compared to the probabilistic approach but its uncertainty envelope is wider than the probabilistic approach, and (iii) the result from Guyonnet et al's algorithms was the most conservative having a very wide level of uncertainty and increased processing time. The analysis result shows the potential benefit of the proposed hybrid approach compared to other approaches.

Moving from one theory to another as appropriate will allow us to utilize the advantages of the both theories (Klir, 1987; Klir and Yuan, 1995; Dubois and Guyonnet, 2011). However, there could be questions on whether to convert all uncertain information into fuzzy or probabilistic form. In this respect, the fundamental principle to be maintained is that no information should be added or lost during a risk assessment or any decision-making process or analysis. Obviously, a fuzzy set theory and membership function is selected when the information is qualitative, imprecise or data is scarce. If the majority of the data is as such, the analysis result will also be less precise and the fuzzy set approach will be the first choice. In contrast, probabilistic approach should be applied, if the majority of the information is available in probabilistic form with few parameters in fuzzy form. As the possibility is an ordinal and epistemic scale of uncertainty, values or figures of possibility are less precise than those of the probability. So, conversion from possibility to probability could be considered as a process to derive less precise probabilities from less precise data sources. In contrast, probability is sensitive to noise and errors. It is well known that small errors of prior probabilities may produce wrong reasoning results. Therefore, the sources of data, numbers of data, level of uncertainty and the objectives of analysis will drive the choice between probability and fuzzy approach.

In this Chapter algorithms for uncertainty modelling were also developed. A simplified flow diagram is presented as a guide to select techniques of uncertainty propagation. The uncertainty propagation techniques that are selected for the analysis are MCS, LHS, bootstrapping and fuzzy alpha-cut technique. The developed uncertainty modelling techniques have been discussed with a simple case example and applied in the case of water demand modelling.

In addition, this Chapter also presents a fuzzy set theory based risk modelling technique. In this method, the likelihood of risk can be either probabilistic (if the likelihood is a model output) or possibilistic (if the likelihood is based on experts' view) and consequences or severity of the risk is represented by fuzzy membership functions. This is because in most of the cases the analysis of severity is a multi-dimensional problem; risk perception and aversion are often difficult to predict by a system model. We expect to get qualitative results rather than precise quantitative results in such cases. A seven-tuple of fuzzy membership functions is considered for both the likelihood and consequences. The magnitude of the risk is calculated by fuzzy alpha-cut technique. For the simplicity and ease of processing, a triangular shaped membership function has been considered. Unlike, likelihood and severity, only a five-tuple of risk level is proposed. The developed risk model has been demonstrated in the Birmingham case study.

The application example used for the demonstration of the developed methodologies is related to the future water demand modelling in Birmingham for year 2035. The future water demand in Birmingham was analysed considering the potential future change pressures with uncertainty, such as population growth, climate change, deterioration of infrastructure (i.e., resulting leakage), micro demand of water and the result was compared to the STWL result.

As presented earlier, changes in future average seasonal temperatures and precipitation rates were calculated in this analysis using the projection proposed by the UKCIP02 Hadley Global Climate Model and the moving block-bootstrapping technique. The relations between climatic parameters and water demand were established using regression modelling. To forecast the future population growth, the moving average method and Monte Carlo simulation technique of uncertainty analysis were both used. Impacts related to socio-economic structures, housing types, and occupancy figures were accounted for using the ACORN classification system. Finally, the micro-demands of water use were analysed based on the records provided by STWL and by using a Latin Hypercube sampling of 1000 runs.

The future per capita water demand was forecasted based on the current patterns in consideration of the potential range of micro-demands planned by OFWAT, UK. The calculated per capita demand was lower than the actual current water consumption rate (i.e., 114.36 LPCD < 141.37 LPCD). Regression models on temperature and precipitation showed that there would be a 2.10% increase and a 0.06% reduction in total monthly water demand if temperatures increased by 1° C and rainfall increased by 1mm/month.

According to STWL, the existing safe water resource capacity for Birmingham is 314.90 ML/day. The forecasted water demands showed that the future state of the Birmingham water source could be characterised within a three-point uncertainty envelope, which predicts that the source will be able to meet future demands until 2022, if future water demand increases at the average ate. It is clear that by 2035, the daily water demand will not be met by the existing source. The calculation also showed that the water shortage rates in Birmingham will be 2.58% (i.e., 8.15 ML/day), and 14.24% (i.e., 44.85 ML/day), for the mean, and maximum growth projections of the uncertainty envelope, respectively. Considering the future daily water consumption rate of 114.36 LPCD, about 71,275 people will be underserved.

This example demonstrated the hybrid approach of uncertainty analysis and risk assessment developed by this research. The result is slightly different from the results of STWL analysis. The presented analysis showed the total water demand; however, sector specific analysis was not possible due to unavailability of sufficient sector-specific data (i.e., domestic, industrial, and commercial). In addition, monthly time steps and average climate characteristics were used for this analysis, which do not account for droughts or other extreme events. Thus, the analysis should be further strengthened to reflect the impacts of floods and droughts as well.

Chapter 4   Multi-criteria Decision-Making Framework under Risk and Uncertainty

The objective of this Chapter is to develop a multi-criteria based decision-making framework for performance assessment and risk management in UWS. The Chapter begins by introducing the basic concept of using a Multi-Criteria Decision Analysis (MCDA) for decision-making in complex systems. It then presents a critical review of commonly used MCDA methods for evaluating UWS performance assessment and risk management. Based on the review, it proposes a fuzzy set theory based MCDA method developed in this research. The newly developed framework and methods are demonstrated through a performance assessment and strategic planning of main urban infrastructure systems in Kathmandu, Nepal. Finally, the developed methodologies and analysis results are discussed and some conclusions are drawn. Part of the discussion and example application are based on Khatri et al. (2011); Khatri and Vairavamoorthy (2013c).

4.1   Introduction

Conventionally, simple and relatively unimportant decisions are made using a single main criterion called single criterion decision-making. However, the decision-making process used in complex infrastructure systems that are based on system performances and risk management need analysis of multiple criteria in economic, social, and technical dimensions. The level of complexity in future decision-making increases due to associated uncertainties within multiple criteria.

MCDA is the process of reaching a decision through the consideration of available alternatives, guided by various measures, rules, and standards called criteria (Zeleny, 1981). The MCDA tools have been used to deal with decision problems in numerous disciplines including social (Stirling, 2006), economical (Zopounidis and Doumpos, 2002), political (Saaty and Cho, 2001), engineering (Hajkowicz and Collins, 2007) and environmental (Pohekar and Ramachandran, 2004). Criteria can be either quantitative that are clear and easy-to-measure, such as dimensions, quantity and price, or qualitative and difficult-to-measure, such as degree of satisfaction, unclear data, colour, and taste (Kassab, 2006). Thus, MCDM may require the application of various theories to capture all the necessary information (e.g., probabilistic, fuzzy). Even when criteria are quantitative type, conflict may arise amongst decision makers concerning their priorities and preferences with respect to the criteria. Thus, it becomes difficult, if not impossible, to solve decision-making problems using a MCDA process without the consideration of appropriate frameworks, right techniques, multiple conflicting factors and the corresponding trade-offs. The following section provides a general overview on structure and process of MCDA.

4.1.1 Multi-criteria decision-making process

The number of steps in a MCDA process varies according to the objective of the analysis (Pohekar and Ramachandran, 2004; Ananda and Herath, 2009; Awasthi et al., 2010). The MCDA process involves the definition of objectives, arranging them into criteria, identifying possible alternatives, and then measure the consequences. A consequence is a direct measurement of the success of an alternative according to given criterion (e.g., cost in dollars and volume in cubic metres) (Chen et al., 2008). Figure 4.1(a) shows the structure of an MCDA problem, where $A = \{A_1, A_2, ..., A_i, ..., A_n\}$ is the set of alternatives, and $Q = \{1_1, 2, .., j, ..., q\}$ is the set of criteria with n alternatives and q criteria. The basic steps required in the decision-making process, in general, are shown in Figure 4.1 (b).

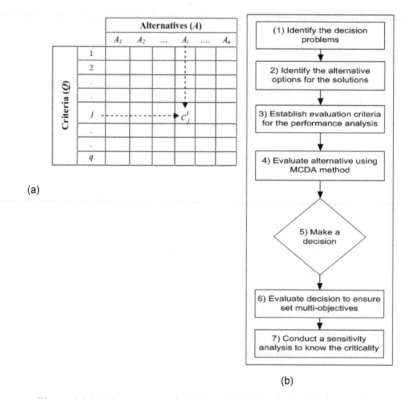

(a)

(b)

Figure 4.1 (a) The structure of MCDA; (b) Multi criteria decision-making process

As shown, the *first step* in a MCDM process is the identification of decision problems that assist in determining the analysis objectives. Based on analysis types, problem identification may include brainstorming, a nominal group technique, a research survey, and potentially a Delphi method (Hasson et al., 2000). Decision options are usually finite, which are ranked or scored creating a 'discrete' choice problem. In some cases, the aim could be to identify an optimum quantity that is subject to constraints creating a 'continuous' choice problem. For example, the decision problem for a water supply system could be the selection of reliable and cost effective water resources to satisfy demand; fixing the price of water for different users; or prioritizing infrastructure investments.

The *second step* is to identify alternatives or options. Alternatives are often formulated according to the analysis objectives. For example, in a water resource augmentation problem, the alternative could be groundwater, rainwater, greywater surface water. In this example, the available alternative could be augmentation by surface water, groundwater, reclaimed water, or other water demand management measures.

The *third step* consists of the identification of main criteria and sub-criteria for the performance measurement and the assignment of relative weights. The selected criteria are mostly represented hierarchically. Some of the qualities to be met by the criteria include completeness, redundancy, operationality, and mutual independence of preferences (Keeney, 1996). The criteria generally selected are multi-dimensional and mutli-sectoral, such as institutional, economic, social, technical, environmental, etc. Additionally, this step requires the involvement of experts, decision-makers and stakeholders.

The *fourth step* consists of the assessment of alternatives when applying the selected MCDA methodology. Generally, a consistent numerical scale for the assessment of criteria is established such that a higher level of performance leads to a higher score value. For example, a scale could assign a score of 0 (worst level) to 100 (best level). Other methods, such as the Analytic Hierarchical Process (AHP) approach of pairwise assessments expresses a judgment of the performance of each alternative relative to each of the other alternatives (Saaty, 1988). Elimination methods rank the decision criteria in an order of preference without using quantitative weights. In the analysis, the preference expression could be weights based (based on criteria) or values based (based on consequences) (Chen et al., 2007). If the preference expression is consequences based, a performance matrix is commonly used. Performance matrices are a standard feature of multi-decision criteria analysis. Additionally, a consequence table could be used, in which each column describes an option or alternative and each row gives the performance of the alternatives against each criterion. The individual performance assessments are often numerical but may also be expressed qualitatively as well (see, Chen et al., 2007).

The final *steps* consist of making a decision based on analysis results. However, before making a final decision, the method should be reassessed to ensure that all essential objectives, criteria, and preferences of the decision makers (DM) have been logically reflected in the decision. At the same time, it is important to analyse the sensitivity of the results based on the different criteria. A systematic sensitivity analysis that will help monitor and implement the plan. Additionally, the sensitivity analysis will reveal where the MCA model needs strengthening and the robustness of result according to the given input assumptions (Hajkowicz and Collins, 2007).

4.2 A Review on Existing MCDA Methods

Recent reviews on MCDA show the growing trend of its application in numerous fields. For example, Romero and Rehman (1987) review 150 MCDA applications in natural resource management; Steuer and Na (2003) review 265 MCDA studies in the field of financial decision-making; Pohekar and Ramachandran (2004) review over 90 MCDA studies in the field of energy planning. Similarly, Hajkowicz and Higgins (2008) review 113 published water management MCDA studies from 34 countries; Ananda and Herath (2009) review more than 60 MCDA research contributions on forest management and planning; Wang et al. (2009) review methods of various MCDA stages for sustainable energy focusing.

MCDA approaches have been classified in numerous ways. One of the first categorizations makes a distinction between multi-objective decision-making (MODM) and multi-attribute decision-making (MADM) based on the number of alternatives under evaluation (Mendoza and Martins, 2006). MADM methods are designed for selecting discrete alternatives while MODM are more equipped to deal with multi-objective problems. The later method is used when a theoretically infinite number of continuous alternatives are defined by a set of constraints on a vector of decision variables (Belton and Stewart, 2002). The general classification of MCDA methods, according to their range of application, has been grouped into three categories as described below. A more detailed description of these classifications can be found in (Belton and Stewart, 2002; Mendoza and Martins, 2006).

i) *Value measurement models*: In this type, scores are developed initially for each individual criterion, and are then synthesized in order to effect aggregation into higher level preference models. The numerical scores represent the degree to which one decision option may be preferred to another. The value measurement models are commonly used either for choosing strategies or technologies, e.g., Multi-Attribute Value Theory (MAVT) (see Hostmann et al., 2005), Multi-Attribute Utility Theory (MAUT) (Keeney and Raiffa, 1993), and AHP (Saaty, 1988).

ii) *Goal, aspiration and reference level models*: First desirable or satisfactory levels of achievement are established for each criterion. The process then seeks to discover options, which are closest to achieving the desirable goals or aspirations. These types are usually applied in situations where decision makers may find it difficult to express trade-offs or importance weights. The methods are able to describe outcome scenarios, expressed in terms of satisfying aspirations or goals for each criterion, particularly to filter out the most unsuitable alternatives. The methods commonly used in this class are Step Method (STEM) (Benayoun et al., 1971); and Technique for Order Preference by Similarity to Ideal Solutions (TOPSIS) (see Behzadian et al., 2012).

iii) *Outranking models*: Initially, alternative courses of action are compared pairwise in terms of each criterion in order to identify the extent to which a preference is favoured over. In aggregating the model seeks to establish the strength of evidence favouring the selection of one alternative over another. Outranking models focus "on pairwise evaluation of alternatives, identifying incomparability, as well as assessing preferences and indifferences. Commonly used models are Elimination and Choice Translating Reality (ELECTRE) (Roy and Vanderpooten, 1996) and Preference Ranking Organization Method for Enrichment and Evaluation (PROMETHEE) (Brans et al., 1986).

Each of the above MCDM methods could be applied in the context of deterministic, stochastic, and fuzzy methods or through a combination of these. There have been many attempts to integrate MCDA with other analytical methods to develop a hybrid approach. As cited in Ananda and Herath (2009), Kangas et al. (1993) employ a combination of AHP and regression analysis to incorporate expert judgment in estimating a suitability function for wildlife habitats; Kurttila et al. (2000) present a hybrid method that integrates AHP and SWOT (strengths, weaknesses, opportunities and threats) analysis. Mendoza and Prabhu (2005) use a hybrid approach to estimate a sustainability index and integrate this with system dynamics. The emphasis of the study was on policy action demonstrating the capability of MCDM models to be used in a combination with system dynamics models for forest issues. Moreover, a linear goal programming method was used to find a satisfying solution to a decision problem that meets a set of aspiration levels rather than maximizing all objectives (Durbach and Stewart, 2003).

De Marchi et al. (2000) present an application of a multi-criteria analysis to assist the municipality of Troina, Italy to identify potential future water resources. In the case study the NAIDE method (Novel Approach to Imprecise Assessment and Decision Environments) (Munda, 1995) was used as aggregation method that can cope with crisp, stochastic or fuzzy data. They illustrate that the technical multi-criteria analysis method alone is not sufficient to cope with the institutional and qualitative component of the water policy problem in Troina . This was strengthened by including social factors such as institutional analysis, participatory observation and a resident survey. This lead to the opportunity for the community dialogue while analysing the water resources problem in Troina.

Yan and Vairavamoorthy (2003) develop a hierarchical model for assessing the conditions and risk of failures of pipes in water distributions systems using a fuzzy composite programming MCDA method. This model considers several indicators affecting water pipes established from the historical data available in other publications. The relative importance of the indicators is assigned by subjective weighting form the decision makers. This hierarchical structure of physical and environmental indicators was used for fuzzy synthetic programming. After fuzzy synthetic programming, fuzzy ranking was done by using the existing maximizing and minimizing set approach. The uncertainty associated to the input indicators was dealt by fuzzy membership functions.

Abrishamchi et al. (2005) developed a compromise programming MCDA in selecting the best possible alternatives for distribution of both available and the transmitted water in the city of Zahidan in Iran. The study used 9 criteria and 8 alternatives. The results obtained reveal that the method is capable of decision-making in comprehensive urban water management studies provided there is some improvement in institutional aspects.

Haijkowicz and Collins (2007) review the use of multi-criteria analysis in water management in 34 countries. They found that multi-criteria analysis has been used in particular for the evaluation of water policies (about 26%) and the strategic planning of infrastructure projects (about 15%). Based on the review fuzzy set analysis, paired comparison and outranking methods are the most common approaches applied. Hajkowicz and Higgins (2008) compare water resource planning and management problems using five multi-criteria methods: weighted summation, range of value, PROMTHEE, Evamix and compromise programming. Their analysis results show that results differ only slightly regardless of the calculation method used. They emphasize the importance of structuring the decision problem (i.e., identifying decision options, identifying assessment criteria, identifying criteria weights and obtaining performance measures) rather than selecting a suitable multi-criteria analysis method. However, the selected method should be capable of handling the available data.

Paneque Salgado et al. (2009) applied the NAIDE (Novel Approach to Imprecise Assessment and Decision Environments) method (Munda, 1995) based software of MCDA to the field of water governance in Costa del Sol Occidental in Málaga, Spain. The objective was to find out whether this concept and combination of tools might be helpful in implementing the integrated evaluation and participation objectives established in the European Water Frame Directive. The main features of the NAIDE are: (i) capable of handling qualitative, quantitative, precise and fuzzy information, (ii) no differential weighting of the various criteria used to evaluate the alternatives, (iii) the purpose is not to produce an undisputable or "optimum" ranking of alternatives. The paper suggests some methodological considerations for applying participatory multi-criteria evaluation based on the outcomes of the case study application.

Earlier discussion shows that the selection of an appropriate method depends on the type of input data, the purposes of the analysis, the decision-making process, and the area of application (Guitouni and Martel, 1998). The selection of appropriate MCDM methods for a particular type of problem has been widely described in several studies (Roy and Vanderpooten, 1996; Belton and Stewart, 2002;

Hajkowicz and Higgins, 2008; Ananda and Herath, 2009). Some of the major findings from the review can be summarised as below:

The comparative study of Haijkowicz and Collins (2007) find that the results of MCDA differ only slightly when analyzed with simple to complex methods. Thus, they emphasize the importance of structuring the decision problem so that it includes the identification of decision options, the selection of assessment criteria, and the assignment of criteria weights and obtains performance measures rather than to select suitable multi-criteria analysis methods.

The MCDA methods are better suited for participatory decision-making processes. Therefore, the multi-criteria analysis methods should allow stakeholder participation for the identification of the performance criteria, the weighting of the criteria, the formulation of the relationship of the criteria, and the actual decision-making process (Haijkowicz and Collins, 2007; Paneque Salgado et al., 2009). In order to include stakeholders, there needs to be a simpler, easier to understand, and more accessible and transparent MCDA method.

The decision options obtained from MCDA are comparable. The challenges pointed out in most of the discussions are on how to incorporate many of the parameters and describe them appropriately in the decision-making process as the criteria may contain different types of information, including crisp, stochastic, or qualitative types, and in many cases there could be limited data. In response to this, some efforts such as NAIDE, developed by Munda (1995), seem promising; however, it does not allow the assignment of weights to the criteria. Also, it is not clear how it works while taking the view of stakeholders and setting their relative importance to the different attributes.

Mendoza and Martins (2006) emphasise qualitative and linguistic information become more relevant and should be better included in MCDA methods. Fuzzy set based techniques to address linguistic information have been developed by Yan and Vairavamoorthy (2003) and in NAIDE (Munda, 1995). However, it is unclear there methods assess if there is only a single data types, such as only probabilistic information, fuzzy information and crisp. Their studies recognised the importance of uncertainty analysis; however, lacks a discussion on uncertainty analysis techniques. Thus, the challenge is developing a MCDA method that has the capacity to address multiple sources of information, uncertainty analysis, potential involvement of stakeholders while being simple to use and transparent. As pointed out by Roy (1985) "the main aim of multiple criteria decision aid is not to discover a solution, but to construct or create a set of relations amongst actions that better inform the actors taking part in a decision process". Still most of the multi-criteria analysis methods have limitations when dealing with data sets that are uncertain, ambiguous, and linguistic (Silvert, 2000; Roy, 2005). To address these shortcomings, this research proposes a fuzzy set based MCDM method for the performance assessment and risk management of UWS. The proposed MCDM method is presented in the next section.

4.3 Proposed Method of MCDA

4.3.1 Introduction

This research proposes a fuzzy set theory based MCDA method. Fuzzy sets and fuzzy logic are extensively applied and well accepted in decision-making (Zimmermann, 1987), fuzzy control and fuzzy systems (Pedrycz, 1993), forecasting using an adaptive-network-based fuzzy inference system (Jang, 1993), uncertainty analysis (Dubois and Prade, 1980; Zadeh, 1983; Klir, 1990a), Clustering of data from large data set and producing a concise representation of a systems behaviour (Chiu, 1994) and infrastructure management (Flintsch and Chen, 2004). More detailed descriptions of fuzzy sets, fuzzy logic, fuzzy operations and their application are available elsewhere (e.g., (Dubois and Prade, 1980; Kaufmann and Gupta, 1985; Klir and Yuan, 1995).

The design of the fuzzy system requires the definition of a set of membership functions (MFs) and a set of fuzzy rules. Commonly used MFs include triangular, trapezoidal, generalized bell-shaped and Gaussian functions. However, triangular and trapezoidal MFs are most frequently used in fuzzy applications due to their simplicity in expression and computation (Ross, 2008). Fuzzy rule-based reasoning incorporates imprecise and qualitative data in the decision-making process by combining descriptive linguistic rules (Fisher, 2003). A fuzzy *if-then rule* is a statement in which some words are characterized by continuous MFs. For example, *IF the effectiveness is poor AND reliability is good AND cost is satisfactory, THEN the performance of a system is satisfactory.* A brief introduction on fuzzy set theory focusing on uncertainty analysis and application of fuzzy set theory for risk assessment is presented in Chapter 2. Some of the basic fuzzy operation and fuzzy arithmetic used in this thesis is presented in Appendix II.

There are many MCDA methods that are based on the fuzzy set theory (see Awasthi et al., 2010). Fuzzy synthetic evaluation (FSE) technique is one of several fuzzy MCDM methods in which various individual components of an evaluation are synthesized into an aggregated form. Fuzzy classification, fuzzy similarity and fuzzy comprehensive assessment are all examples of the FSE technique (Rajani et al., 2006). It has been widely applied in various fields, such as analysis of urban traffic environment quality (Tao and Xinmiao, 1998); assessment of reservoir water quality (Lu and Lo, 2002); identification of river water quality (Cheng and Chau, 2001); assessment and prediction of the overall environmental quality of air, water and soil in a city (Haiyan, 2002); ranking of pipes in a water distribution network according to their condition ratings (Yan and Vairavamoorthy 2003); evaluation

114

of disinfection by-products using risk-based indexing (Sadiq and Rodriguez, 2004); and condition rating of buried pipes (Ranjani et al., 2006). The following section presents the major steps in the MCDA process proposed from this research.

4.3.2 Major steps of proposed MCDA method

The proposed FSE technique to assess the performance of infrastructure systems and to select a feasible risk management strategy in UWS consists of the following five steps. The following discussion covers the steps for both the performance assessment and risk management. It is assumed that the MCDA method used for the overall system performance assessment is of a larger scale than that used for the risk management (i.e., the framework for the risk assessment will be a subset of all performance assessment framework).

   i)   Identification and classification of performance indicators;

   ii)   Fuzzification of the performance indicators;

   iii)   Weight calculation and weight assignment;

   iv)   Aggregation of performance indicators; and

   v)   Defuzzification of the aggregated indices to produce the overall systems performance.

Identification and classification of performance indicators

Selection of a set of criteria and their corresponding indicators is one the first and important steps of MCDA (see Figure 4.1 b). The performance criteria are the means of measuring the performance for the system, or its parts, by society or utility operator that builds, operates, uses, or is neighboured to that infrastructure. In the context of urban infrastructure systems performance is understood to mean supplying adequate quality and quantity of drinking water, maintaining system reliability, safe movement of people and goods from one point to another, safe removal of wastes, or the safe completion of additional tasks (NRC, 1996).

Most of the cases, the types of performance criteria depend on the purpose of the analysis and the local conditions. For example, Alegre et al. (2006) and Matos et al. (2003) developed performance indicators that were only applied to water supply and wastewater systems; Kolsky and Butler (2002) developed and applied performance indicators to urban storm drainage systems in developing countries; Jasch (2000) developed performance indicators for environmental management; and Talvitie (1999) did the same for the road sector. Other works developed analytical frameworks for measuring performance, such as the study of the Rotterdam drainage system by Geerse and Lobbrecht (2002) and the measurement of the sustainability of a power infrastructure system by Dasgupta and Tam (2005). One of main precautionary suggestions proposed by most of the studies is ensuring a balance is kept between cost, availability of data, scope, complexity, and accuracy during the development of indicators (Milman and Short, 2008).

This research proposes a hierarchical structure of performance criteria into three hierarchical levels: performance indicators, dimensions and categories (Figure 4.2). The performance categories are the broader performance measures of a system and are classified as effectiveness, reliability and system cost. The performance categories are further disaggregated into nine performance dimensions, which are sub-criteria, to reflect different aspects of the system including adequacy, quality and reliability of the service; environmental impact; and the financial, social and economic benefits. Finally, a number of indicators are selected for each dimension. The performance criteria will vary according to the system type, context, time, and economy of a city (Matos et al., 2003). It is noted that the approach used here for characterization and classification of performance indicators resembles that presented by Ashley et al. (2004) for incorporation of sustainability in the UK water industry.

As shown in the framework above (Figure 4.2), the first performance criterion is *effectiveness*, which measures the ability of a system to provide services according to the community's expectations. It refers to the extent to which objectives of programmes, operations, or processes have been achieved. This category consists of four types of performance dimensions: i) *service delivery capacity*; ii) *quality of services*; iii) *regulatory concerns*; and iv) *community concerns*. The service delivery capacity and quality of services dimensions measure the objectives of the systems set by infrastructure design professionals. These dimensions include those indicators that measure adequacy, coverage, daily supply hours and impact on public health, among others. Similarly, the regulatory dimension measures the quality of standards met by a system. Finally, the community concerns dimension measures the externalities that a system should address, such as the overall well-being of the society in terms of the standard of living, equity in services and more.

The *reliability category* measures the ability of a system or component to perform its required functions under stated conditions for a specified period. It is categorized into three dimensions: i) *reliability of service*; ii) *reliability of quality*; and iii) *reliability of infrastructure capacity*. These dimensions include the level of services, quality of services and the capacity of a system over its designed period, respectively. A few examples of reliability indicators categorized under these dimensions are: probability of expected and unexpected system failure, performance of the treatment plants, etc.

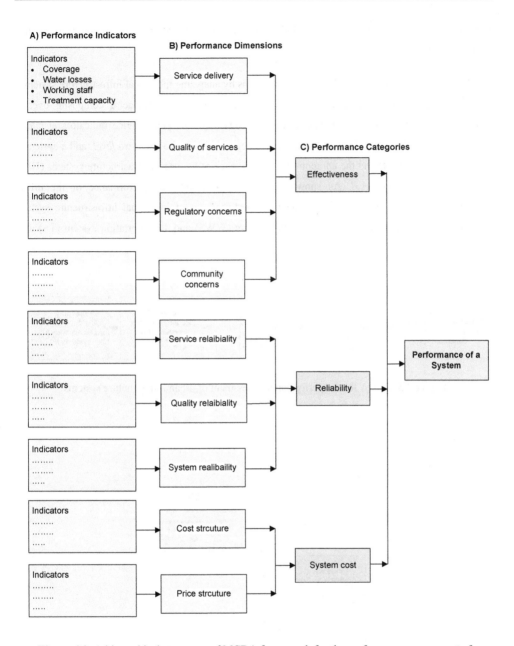

Figure 4.2 A hierarchical structures of MCDA framework for the performance assessment of infrastructure systems (Khatri et al., 2011)

Finally, *system cost* category measures the financial and economic justification of the infrastructure investment. It consists of *cost* and *price* dimensions. The indicators used to measure the cost dimension include costs required for construction, operation and maintenance. Similarly, indicators used to measure the price dimension include tariff structures, the revenues collection rate, and the basis for fixing fees or charges.

The MCDA framework is hierarchical in that it begins by analyzing individual infrastructure systems, generates indices for them, and then combines these indices to produce a composite overall infrastructure index (Figures 4.3). The framework consists of the performance indicators at *micro level* (at level 1), the dimensions (at level 2) and categories (at level 3) at *Meso level*, and a system's performance (at level 4) and the compounded performance of the urban infrastructure systems (UIS) (at level 5) at *Macro level*. As shown in Figure 4.4, aggregated performance of the urban infrastructure systems (UIS) is the compounded performances of individual infrastructure system (e.g., water supply system (WSS), wastewater system (WWS) and Transportation system (TS) and other systems in an urban area). The framework can be applied for a single system or multiple systems in an urban area.

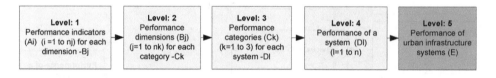

Figure 4.3 MCDA for the performance assessment of urban infrastructure systems

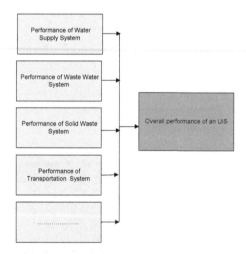

Figure 4.4 MCDA framework for the performance assessment of UWS

MCDA framework for risk management

The objective of risk management is to identify potential solutions or strategies to reduce or control the identified risk. Therefore, the MCDA based risk management method helps the decision maker select options such that risk is reduced or controlled in the future (Kristensena et al., 2006). It will require decision makers to consider the tradeoffs of potential options or strategies with other dimensions, such as environmental, economic, social, cultural, institutional and technical (Haimes, 2009b).

The framework proposed for selecting an option or strategy for the risk management is shown Figure 4.5 (b). It has three hierarchical levels: performance indicators, performance sub-criteria (i.e., performance dimensions) and finally the overall performance. The indicators included within those criteria are, for example, i) *Environmental*: energy demand; environmental impact with respect to minimum flow requirement; potential pollution load to receiving water bodies; nutrients recovery rate etc., ii) *Economic*: life cycle unit cost; willingness to pay for the service etc., iii) *Technical*: space requirements; reliability of services; efficiency of resources; complexities of technology etc., iv) *social and institutional*: acceptability of service; risk to the public health etc. It is noted that the indicators under each of the sub-criteria may vary according the availability of data, purpose of analysis, and context.

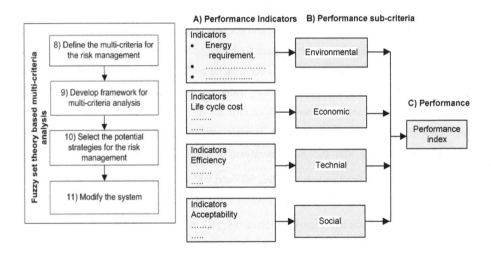

Figure 4.5 (a) A framework for risk management, (b) MCDA for risk management

It is noted that the process of MCDA for risk management in UWS is presented in Chapter 2 and partly in Figure 4.5 (a), where steps 8, 9, 10, and 11 are related to the risk management. Figure 4.5 (b) shows the step 8, where as step 9 uses the framework presented in the proceeding section. Steps 10 and 11 are further illustrated in the Mbale case study presented in Chapter 5.

Fuzzification of the performance indicators

Fuzzification is the process whereby a crisp number is converted to a fuzzy set. The converted fuzzy number is represented by membership functions. As presented in Chapter 3 there are no strict rules or theories for the selection of the number of membership functions. In order to simplify the analysis without losing too much information, the membership functions have been classified into four levels: poor, satisfactory, good and excellent. These quality levels have been represented by the triangular and trapezoidal shaped membership functions. It is noted that the fuzzification technique will depend on the information contained in the performance indicator (e.g., crisp, fuzzy or probabilistic inputs).

In this classification, the *excellent level* indicates the highest possible level of a system's performance that is technically feasible and already achieved in many of the best performing cities around the world. The *good level* indicates a performance that is acceptable for a local condition. Similarly, the *satisfactory level* requires further performance improvement. The performance level below satisfactory is *poor level meaning the systems' performance is* unacceptable given the current conditions and the system requires an immediate plan of action for improvement.

As discussed in Chapter 3 it applies a maximum operator to determine the membership value (Klir and Yuan, 1995). For example, assume an uncertain indicator, A_1, for a system that has three values of measurement (i.e. $X_1 = 25$, 32 and 45 and is represented by the dashed triangle in Figure 4.6) that correspond to the minimum value, the most likely value and the maximum value, respectively. When X_1 is mapped on a fuzzy scale, X_1 intersects with different membership functions that result from the 4-tuple fuzzy set. For example, X_1 does not intersect with the excellent membership function (i.e., $\mu_{excel} = 0$); it intersects with the good membership function at 0.15 (i.e., $\mu_{good} = 0.15$); the satisfactory membership function at two points, 0.75 and 0.85 (thus requires the maximum operator); and it intersects with the poor membership function at 0.31 (i.e., $\mu_{poor} = 0.31$). Thus, after the fuzzy max operation (i.e., with the maximum of 0.75 and 0.85) for the satisfactory range, it will result in the satisfactory fuzzy membership value of 0.85 (i.e., $\mu_{satif} = 0.85$). Thus, the fuzzified 4-tuple fuzzy set for indicator A_1 will be [0.31, 0.85, 0.15, 0]. Similarly, if another indicator, A_2, has a crisp value of measurement (i.e., $X_2 = 55$ and is represented by a dashed line in Figure 4.6), then the fuzzified value will be [0, 0.25, 0.75, 0] after the fuzzy mapping operation. Thus, a 4-tuple fuzzy set generated after the fuzzification operation is generally given by

$$A_i = [\mu_1 \quad \mu_2 \quad \mu_3 \quad \mu_4] \qquad (4.1)$$

where A_i is a 4-tuple fuzzy set (poor, satisfactory, good and excellent) for an indicator, i, after the fuzzification operation.

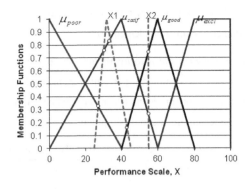

Figure 4.6 Membership functions and fuzzification

Assignment of weights

The relative importance of a performance indicator in MCDA is represented by assigning weights. However, the assignment of weights of a performance indicator, in most cases, is a subjective exercise. As it is usually assigned based on consultation with decision-makers and experts. Thus, the assigned weights will be influenced by the knowledge of those consulted and by their preferences and biases. Analyzing the preferences, biases and their possible treatments is outside the scope of this work and further detailed analysis in this respect can be found elsewhere (see, (Gilovich et al., 2002; Keeney, 2002; Jacobi and Hobbs, 2005; Hämäläinen and Alaja, 2008)).

There are various subjective weighting methods in literature that can be applied to the MCDM techniques, including pairwise comparison, the simple attribute rating technique (SMART), Swing, AHP and more (see, Wang et al., 2009). In this paper, the AHP technique has been applied as it is widely used and has been reported to be a simple but very effective subjective weighting method (Zahedi, 1986; Lu et al., 1999; Sadiq and Rodriguez, 2004). The AHP technique involves the following three steps.

i) Pairwise comparison of the performance indicators and the development of a judgment matrix: The relative importance of the performance indicators is judged by a pairwise comparison of the indicators. The experts express their preferences for each pair of the set, considering the importance of one indicator against the other by ranking how much or by how many times they prefer the indicator. Saaty's intensity scale is used for ranking the preferences in nine levels. They are equally preferred (assigned value of 1), weakly preferred (assigned value of 3), moderately preferred (assigned value of 5), strongly preferred (assigned value of 7), or absolutely preferred (assigned value of 9). The intermediate values within each level are translated with 2, 4, 6 and 8

121

respectively (see (Saaty, 1988). A judgment ratio matrix of size: $n \times n$ will results based on the decision made on the number of compared elements (n).

ii) *Synthesis of the matrix to compute the priority vector:* The judgment matrix is synthesized to calculate the priority vector. It is commonly performed using mathematical techniques such as Eigenvectors, mean transformation or geometric mean (Zahedi, 1986). In this case, we have applied the geometric mean technique due to its simplicity.

iii) *Checking the consistency of the judgment matrix:* Finally, the level of inconsistency in the judgment matrix is checked by computing the consistency ratio as

$$CR = (CI / RCI) * 100 \tag{4.2}$$

where *CI* is the *consistency index* (Equation 4.3) and *RCI* is the *random consistency index* (Equation 4.4).

$$CI = (\lambda_{max} - n)/(n-1) \tag{4.3}$$

$$RCI = (\overline{\lambda}_{max} - n)/(n-1) \tag{4.4}$$

where λ_{max} is the maximum Eigenvalue of the judgment matrix, n is the dimension of the pairwise comparison judgment matrix, and λ_{max} is the average Eigenvalue of the judgment matrix derived from randomly generated reciprocal matrices using the scale 1/9, 1/8,...1/2, 1, 2....8, 9 and the pairwise comparison for a very large sample. Table 4.1 presents the RCI values used to check the inconsistency in the AHP judgment matrix.

Table 4.1 Average random consistency index (RCI) table

Matrix size:	1	2	3	4	5	6	7	8	9	10
RCI:	0	0	0.58	0.90	1.12	1.24	1.32	1.41	1.45	1.49

The standard rule recommended by Saaty (1988) indicates that the consistency ratio should be less than or equal to 10% for decision-makers to be consistent in their pairwise judgments. Saaty (1988) has also shown that is always greater than or equal to n. Thus, the closer the value of computed λ_{max} is to n, the more consistent the observed values of matrix are. If the CR is larger than desired, it is treated in three ways (Saaty, 1988): (i) by finding the most inconsistent judgment matrix; (ii) by determining the range of values to which that judgment matrix can be changed so that the inconsistency would be improved; and (iii) by changing the judgment matrix by making revisions and improvements in a plausible range.

The relative importance of the performance indicators is calculated by using the aforementioned steps of the AHP. Further illustration of the AHP application is presented in the application section.

Aggregating of performance indicators

The aggregation operation consists of synthesizing the lower level performances to the upper levels. As shown in Figure 4.3, the compounded performance of indicators at level 1 will result in the performance dimension, and these performance dimensions at level 2 will result in the performance category. Subsequently, performance categories combined at level 3 will give the system performance. Finally, the performances of all the systems are synthesized at level 4 to calculate the overall performance of UWS (at level 5). The resulting synthesized performance value at any level will be a 4-tuple fuzzy set.

The aggregation operation at any level is performed by a matrix operation of the vector of the weights of the indicators and evaluation matrix. For example, let there be n_j indicators (at level 1) under a performance dimension, B_j (at level 2). After the fuzzification of n_j indicators, the result will be an evaluation matrix of size $[n_j * 4]$; then, if a vector of the weights of the indicators is given by, $W_{jn_j} = [w_{j1} \ ... \ w_{ji} \ ..... w_{j\,n_j}]$ then the aggregation operation will be given by

$$B_j = W_{jn_j} * A_{j\,n_j} \tag{4.5}$$

$$B_j = [w_{j1} \ ... \ w_{ji} \ ..... w_{j\,n_j}] * \begin{bmatrix} \mu_1^{j1} & \mu_2^{j1} & \mu_3^{j1} & \mu_4^{j1} \\ \vdots & \vdots & \vdots & \vdots \\ \mu_1^{ji} & \mu_2^{ji} & \mu_3^{ji} & \mu_4^{ji} \\ \vdots & \vdots & \vdots & \vdots \\ \mu_1^{jn_j} & \mu_2^{jn_j} & \mu_3^{jn_j} & \mu_4^{jn_j} \end{bmatrix} \tag{4.6}$$

$$B_j = [\mu_1^j \ \ \mu_2^j \ \ \mu_3^j \ \ \mu_4^j] \tag{4.7}$$

where $A_{j\,n_j}$ is an assessment matrix of dimensions n_j by 4 formed after the fuzzification of n_j indicators under a dimension j; $(\mu_1^{ji}, \ \mu_2^{ji}, \ \mu_3^{ji}, \mu_4^{ji})$ is a 4-tuple fuzzy set (poor, satisfactory, good, and excellent level) for an indicator, i, $(i = 1 \ to \ n_j)$; and B_j is a 4-tuple fuzzy set $(\mu_1^j, \ \mu_2^j, \ \mu_3^j, \mu_4^j)$ for a dimension after the aggregation.

The same operations (Equations: 4.5, 4.6, and 4.7) are repeated for the subsequent upper level 3, 4 and 5, respectively. The synthesized performance result, at level 4, will be the performance of each infrastructure system, and the subsequent synthesized performances of all the individual systems, at level 5, will be the overall performance of the urban infrastructure system.

Defuzzifying of the aggregated indices to produce the overall systems performance

In this step, a 4-tuple fuzzy set is converted to a crisp number by a defuzzification operation. The defuzzification can be performed with several techniques, including the centroid method, mean-max-membership operation, a maximum operator, first of maximum, last of maximum and mean of maximum and by assigning weights to its membership function (Silvert, 2000). One of the commonly applied techniques for defuzzification is the centroid method and it has been applied in this research. A centroid defuzzification method finds a point representing the centre of gravity of the fuzzy set and is given by

$$C_A = \int_a^b \mu_A(x)x\,dx \Bigg/ \int_a^b \mu_A(x)$$

(4.8)

where C_A is the centroid of a fuzzy set, A, on the interval, ab.

The defuzzification operations will be undertaken at levels 4 and 5 (Figure 4.3). As mentioned, the corresponding defuzzified results will be the PI of a single system and overall UIS, respectively. We have selected the four quality levels to represent the performance of a system. Therefore, the centroid values of each membership function will be considered during the PI calculation and are given by

$$Index_l = [D_l] \bullet [C]^T$$

(4.9)

where D_l is the performance of a system, l, in a 4-tuple fuzzy set form, C^T is a transpose of a vector of centroid values of the membership functions, $C = [C_{poor} \quad C_{satif} \quad C_{good} \quad C_{excl}]$, and *Index* is the performance index (PI) of a system, l.

Similarly, the overall performance $Index_{(overall)}$ of a UIS is calculated by defuzzifying the 4-tuple fuzzy set of the overall system's performance *(E)* and is given by

$$Index_{(overall)} = [E] \bullet [C]^T$$

(4.10)

where E is the 4-tuple fuzzy set of aggregated performances of all the systems under consideration.

4.3.3 Summary of the method

A new MCDA method based on the fuzzy set theory named as *Fuzzy Synthetic Evaluation (FSE)* method has been proposed from this research. An algorithm to assess the performance of an infrastructure system using aforementioned framework and methods (Figures: 4.1, 4.2, and 4.3, and Equations: 4.1 to 4.10) is presented in Figure 4.7. A computational tool (in MATLAB) was developed to analyse the performance of the system based on the presented frameworks and algorithm. It is noted that the flow diagram (Figure 4.7) is applicable for a single system. The integrated performance of entire systems or multiple systems can be calculated by aggregating the performance of independent systems.

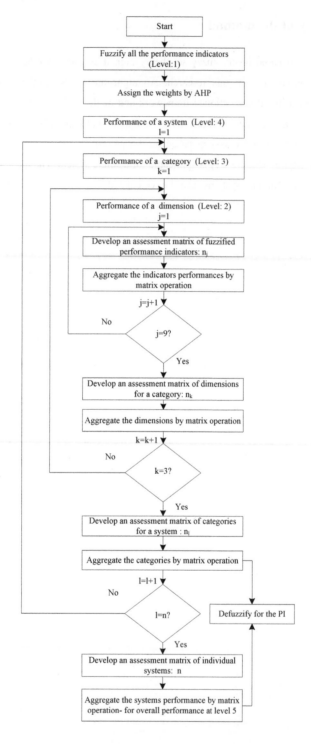

Figure 4.7 Flow chart for computing the overall performance of an UIS

4.4 Application of MCDA for Performance Assessment of Water Supply System in KMC, Nepal

The developed MCDA framework, FSE technique and the computational tool was applied to analyze the performance of four civil infrastructure systems, namely water supply (WSS), wastewater (WWS), transportation (TS) and solid waste (SWS), in Kathmandu metropolitan city (KMC) Nepal. The wastewater system includes sewerage and urban drainage system whereas urban road and traffic systems are covered in transportation system. Using the performance results, recommendations are made for the strategic planning of infrastructure systems in KMC. Additional information on the case study is also available in Khatri et al., (2011). The subsequent section presents the details of the analysis.

4.4.1 Introduction

KMC is the capital city of Nepal with a population around 1 million (KMC Office, 2010). The city spreads over an area of 50.67 km^2, and is divided into 35 administrative wards. Most of the country's ministries, central departments, universities, major hospitals, and business centre are situated within KMC. Various governmental, semi-governmental, and private sector institutions are involved in managing the infrastructure system in KMC. For example, the Nepal Water Supply and Sewerage Corporation (NWSC) is responsible for drinking water supply and wastewater treatment services; the city roads are managed by the KMC and the Department of Roads (DOR); and solid waste management is the responsibility of the Ministry of Local Development (MOLD) and the KMC.

In order to apply the tool to KMC, data were collected in cooperation with the various organizations described above. Five consultation meetings were held with experts from these organizations (i.e., engineers, consultants, planners, managers and accountants) to identify the type of indicators to be incorporated into the modelling framework. In addition, three workshops were held with the experts to establish the relative importance of various performance indicators; performance dimensions; performance categories; and different infrastructure systems. The outcomes from these workshops were used to generate weights for all stages of the aggregation process. After developing the modelling framework, a final workshop was undertaken to refine the framework and to establish consensus in relation to the weightings (that reflect their relative importance of sleeted criteria).

In order to apply the tool to KMC, data was collected in cooperation with the various organizations described above. Five consultation meetings were held with experts from these organizations (i.e., engineers, consultants, planners, managers and accountants) to identify the type of indicators to be incorporated into the modelling framework. In addition, three workshops were held with the experts to establish the relative importance of various performance indicators; performance dimensions;

performance categories; and different infrastructure systems. The outcomes from these workshops were used to generate weights for all stages of the aggregation process. After developing the modelling framework, a final workshop was undertaken to refine the framework and to establish consensus in relation to the weightings (that reflect their relative importance of sleeted criteria).

4.4.2 Performance indicators and fuzzification

The types of performance indicators incorporated in the WSS are presented in Table 4.2. The data for the selected indicators was collected from primary and secondary sources (mainly published and unpublished reports), and from detailed discussions with the relevant experts. Where fuzzy indicators were involved, four levels of classification were defined (i.e., poor, satisfactory, good and excellent), and these were established through detailed discussions with experts. For example, one of the indicators used for water supply systems is unaccounted for water (UFW). The range used for UFW was established from relevant literature (Tortajada, 2006), which states that the most efficient urban water systems operate with a UFW of less than 10% (e.g., Singapore: 6% and Japan: 8%), while the worst operate at UFW rates of over 50%. This information was shared with the KMC experts, who then developed the following scale for UFW: excellent, if $UFW \leq 10\%$; good, if $10\% \leq UFW \leq 15\%$; satisfactory, if $15\% \leq UFW \leq 25\%$; and poor, if $UFW \geq 25\%$. Table 4.2 presents the classification of other indicators used in the effectiveness category of the WSS that were established in a similar fashion.

Table 4.2 Selected MCDA criteria for water supply system with the observed filed data for the respective indicators and calculated weights after AHP operation

Performance Categories/Performance Dimensions/ Description of Indicators	Observed Field Values	Assigned Weight
1) Effectiveness		0.33
A) Dimensions Type: Service Delivery		0.34
• Adequacy (Total demand met by the system): %	55%	0.31
• Coverage (Population covered): %	78%	0.31
• Water losses (UFW): %	37%	0.19
• Working staff: Numbers/1000 connections	9.6	0.09
• Treatment plant capacity with respect to maximum demand: %	55%	0.10
B) Dimensions Type: Quality of Services		0.29

• Average daily supply hours: Hrs/day	2	0.32
• Customers cares: % of population satisfied	30%	0.28
• Fire demand: % of fire demand met by the system	30%	0.10
• Impact on public health (Cholera outbreak incidents numbers in last 10 years)	3	0.30
• C) Dimensions Type: Regulatory Concerns		0.18
• Water quality standard: Compared to WHO	G	0.35
• New connection time: Days	S	0.25
• Accountability: Existing rules/regulations and application	P	0.26
• Water metering: Total % of water metered	83%	0.14
• D) Dimensions Type: Community Concerns		0.18
• Standard of living: % of population with indoor plumbing	E	0.30
• Equity: % of poor with water access	P	0.16
• Environmental sustainability considering water demand management, water table protection, sludge removal practice	S	0.24
• Willingness to pay for the existing tariff :%	E	0.30
2) Reliability		0.33
• A) Dimensions Type: Service Delivery Reliability		0.31
• Unplanned main line failures: Hours/year	12	0.36
• Unplanned distribution system failures: Hours/year	36	0.25
• Unplanned failures due to other CIS: Hours/year	6	0.14
• Unplanned power / pump failure: Hours/year	6	0.25
B) Dimensions Type: Quality Reliability		0.28
• Probability of source pollution (Point & non point pollution)	S	0.30
• Supply system pollution: In and ex-infiltration	P	0.40
• Treatment plant performance: Pathogen removal	S	0.30
C) Dimensions Type: System Reliability		0.41

• Source capacity (Adequacy to design period): %	75	0.31
• Supply system (Capacity of main): %	50	0.28
• Treatment plant (Capacity of treatment plant): %	50	0.31
• Age of distribution system (Average service life): years	25	0.10
3) System Cost		0.33
A) Dimensions Type: Cost Structure		0.5
• O & M Cost (Compared to planned): %	G	0.20
• Current ratio: Current assets/current liabilities	0.50	0.31
• Operating ratio: Operating cost/revenue	0.50	0.35
• Monthly water cost with average monthly income: %	0.50	0.14
B) Dimensions Type: Price Structure		0.5
• Unit tariff : $/m3	0.20	0.41
• Tariff structure (considering flat, or block, or adjusted bloc rate)	G	0.3
• Criteria to fix fees/charges for new connection (considering quantity, distance, pressures)	S	0.15
• Monthly revenue collection rate in time: %	90	0.14
[Note: P= Poor, S= Satisfactory, G= Good, and E= Excellent]		

Table 4.3 Characterization of the performance indicators for the effectiveness category of the WSS

Performance Measures	Performance Quality Levels			
	Excellent	Good	Satisfactory	Poor
1) Effectiveness				
A) Dimensions Type: Service Delivery				
i) Adequacy (Total demand met by the system): %	>80%	60% - 80%	40% -60%	<40%
ii) Coverage (Population covered): %	>80%	60% -80%	40% -60%	<40%
iii) Water losses (UFW): %	<10%	10% -15%	15% -25%	>25%
iv) Working staff: Numbers/1000 connections	<2	2-4	4 -10	>10
v) Treatment plant capacity with respect to	>100%	90% -100%	80% -90%	<80%

	maximum demand: %				

B) Dimensions Type: Quality of Services

i)	Average daily supply hours: Hrs/day	>12 hours	8 - 12 hours	4 -8 hours	<4 hours
ii)	Customers cares: % of population satisfied	>80%	60% -80%	40% -60%	<40%
iii)	Fire demand: % of fire demand met by the system	>80%	60% -80%	40% -60%	<40%
iv)	Impact on public health (Cholera outbreak incidents numbers in last 10 years)	Nil	0 - 2 times	2 - 4 times	>4 times

C) Dimensions Type: Regulatory Concerns

i)	Water quality standard: Compared to WHO	>WHO	= WHO	<WHO	No
ii)	New connection time: Days	< 3 days	3 - 5 days	5 - 7 days	>7 days
iii)	Accountability: Existing rules/regulations and application	>80%	60% - 80%	40% - 60%	>40%
iv)	Water metering: Total % of water metered	> 80%	60% - 80%	40% - 60%	<40%

D) Dimensions Type: Community Concerns

i)	Standard of living: % of population with indoor plumbing	>80%	60% - 80%	40% - 60%	<40%
ii)	Equity: % of poor with water access	>80%	60% - 80%	40% - 60%	<40%
iii)	Environmental sustainability considering water demand management, water table protection, sludge removal practice (% of average criteria met)	>80%	60% - 80%	40% - 60%	<40%
iv)	Willingness to pay for the existing tariff : % of total population	>80%	60% - 80%	40% - 60%	<40%

4.4.3 Assignment of weights

As stated above, the relative importance of the various indicators is articulated through weights. These weights were established during workshops with relevant experts where the AHP technique was applied. During these workshops, experts were asked to undertake a series of pairwise comparisons of the various performance indicators and to assign numerical values using the Saaty scale. In addition, the experts were asked to undertake a series of pairwise comparisons of performance dimensions and categories (again using the Saaty scale). Table 4.4 presents the AHP technique used to generate weights for this application, the judgment matrix and priority vector for five indicators.

Table 4.4 The judgment matrix for the service delivery dimension of the WSS

	Indicators	Adequacy	Coverage	Water Losses	Working Staff	T. Plant Capacity	Priority Vector
D =	Adequacy	1.0	1.0	2.0	3.0	3.0	0.31
	Coverage	1.0	1.0	2.0	3.0	3.0	0.31
	Water Losses	0.50	0.50	1.0	3.0	2.0	0.19
	Working staff	0.33	0.33	0.33	1.0	1.0	0.09
	T. Plant Capacity	0.33	0.33	0.50	1.0	1.0	0.10
	Column Sum	3.17	3.17	5.83	11.0	10.0	1.00

Table 4.4 indicates that the adequacy indicator (D_{11}) is equally important as coverage (D_{12}); two times more important than the rate of water losses (D_{13}) ; and three times more important than the working staff ratio (D_{14}) and treatment plant capacity (D_{15}), respectively. It is noted that the comparison is modified from Saaty's scale that shows direct comparison between each pair; however, it follows the same principles and techniques for judgement matrix development. The priority vector of the judgment matrix (the last column in Table 4.4) is the normalized geometric mean of the respective rows of the judgment matrix (e.g. $W_1 = 1.78/5.73 = 0.31$), and represents the vector of weights, $W = [W_1, W_2, ..., W_N]$. Finally, the consistency of the judgment matrix is calculated by employing Equations 4.2, 4.3 and 4.4. In this example, the consistency ratio (CR) is 1.39%, which is less than 10%, so the judgment matrix is consistent. Table 4.5 presents the CRs of the judgment matrices for all of the indicators and dimensions for the WSS (see Appendix II for details).

Table 4.5 Consistency test results for the indicators and dimensions of the WSS

No.	Performance indicators	Consistency results
1	Indicators for service delivery dimension	n=5, λ_{max} =5.062, CI= 0.0155, RCI=1.12, CR=1.39 %< 10%
2	Indicators for quality of Services dimension	n=4, λ_{max} =4.019, CI= 0.0063, RCI=0.90, CR=0.70 %< 10%
3	Indicators for regulatory concerns dimension	n=4, λ_{max} =4.066, CI= 0.0063, RCI=0.90, CR=2.44 %< 10%
4	Indicators for community concerns	n=4, λ_{max} =4.011, CI= 0.0035, RCI=0.90,

	dimension	CR=0.40 %< 10%
5	Indicators for service delivery reliability dimension	n=4, λ_{max} =4.077, CI= 0.026, RCI=0.90, CR=2.85 %< 10%
6	Indicators for quality reliability dimension	n=3, λ_{max} =3.002, CI= 0.001, RCI=0.58, CR=0.70 %< 10%
7	Indicators for system reliability dimension	n=4, λ_{max} =4.022, CI= 0.007, RCI=0.90, CR=0.80 %< 10%
8	Indicators for cost structure dimension	n=4, λ_{max} =4.102, CI= 0.034, RCI=0.90, CR=3.80 %< 10%
9	Indicators for pricing structure dimension	n=4, λ_{max} =4.004, CI= 0.001, RCI=0.90, CR=0.10 %< 10%
10	Dimensions for effectiveness category	n=4, λ_{max} =4.012, CI= 0.004, RCI=0.90, CR=0.40 %< 10%
11	Dimensions for reliability category	n=3, λ_{max} =3.001, CI= 0.001, RCI=0.58, CR=0.15 %< 10%
12	Dimensions for system cost category	n=2, λ_{max} =2

4.4.4 Aggregation and defuzzification

After fuzzification and assignment of weights for all the indicators, the aggregation operation was undertaken at levels 2, 3, 4 and 5 using Equations 4.5, 4.6 and 4.7 respectively. The resulting aggregated performance will form a 4-tuple fuzzy set, which was converted into the PI by a defuzzification operation (Equations 4.8, 4.9 and 4.10) at level 4 and 5 in the framework. As mentioned, the triangular membership functions are used to represent the poor, satisfactory and good quality levels (i.e., poor (0, 0, 40), satisfactory (0, 40, 60) and good (40, 60, 80)) and trapezoidal membership functions for the excellent level (i.e., 60, 80, 100, 100). Thus, the resulting centroid values to be used for the analysis are: C_{poor} =13, C_{satf} =33, C_{good} =60 and C_{excel} =85, respectively.

4.4.5 Results and discussion

The developed tool described above was used to analyze the performance of UIS in KMC. Table 4.6 presents the performance results for the WSS, where column 1 presents the fuzzified results for each of the selected performance indicators, and columns 2 and 3 present the 4-tuple fuzzy set (aggregated performance results) of the performance dimension and category. The 4-tuple fuzzy sets indicate the degree to which the performance is excellent, good, satisfactory and poor, respectively. For example, [0.0 0.0 0.6 0.4] indicates: excellent and good are zero; satisfactory is 60%; and poor is 40%. Finally, column 4 is the calculated performance index of the WSS.

Table 4.6 Performance analysis of the WSS

Performance indicators /Performance dimensions		Performance categories		PI of the WSS
(Fuzzification process)	(Aggregation process)	(Aggregation process))		(Defuzzification)
Service delivery				
[0.0 0.75 0.25 0.0]				
[0.90 0.10 0.0 0.0]	[0.27 0.25 0.15 0.33]			
[0.0 0.0 0.0 1.0]				
[0.0 0.0 0.07 0.93]				
Quality of services				
[0.0 0.0 0.5 0.5]				
[0.0 0.0 0.75 .25]	[0.0 0.0 0.60 0.40]			
[0.0 0.0 0.75 0.25]				
[0.0 0.0 0.50 0.50]		Effectiveness	[0.22 0.18 0.33 0.27]	
Regulatory concerns				
[0.0 1.0 0.0 0.0]				
[0.0 0.0 0.0 1.0]	[0.20 0.50 0.10 0.20]			
[0.0 0.5 0.5 0.0]			44.15	
[1.0 0.0 0.0 0.0]			(Satisfactory)	
Community concerns				
[1.0 0.0 0.0 0.0]				
[0.0 0.0 0.75 0.25]	[0.50 0.05 0.40 .05]			
[0.0 0.15 0.85 0.0]				
[1.0 0.0 0.0 0.0]				
Service reliability				45.03
[0.0 0.0 0.50 0.50]				
[0.0 0.0 0.0 1.0]	[0.0 0.20 0.35 0.45]			
[0.0 0.50 0.5 0.0]				(Satisfactory)
[0.0 0.50 0.50 0.0]				
Quality reliability		Reliability	[0.09 0.17 0.42 0.32]	
[0.0 0.0 1.0 0.0]				
[0.0 0.0 0.0 1.0]	[0.0 0.07 0.53 0.40]			
[0.0 0.25 0.75 0.0]				
System reliability			36.03	

[0.75 0.25 0.0 0.0]			(Poor)	
[0.0 0.50 0.50 0.0]	[0.22 0.22 0.40 0.15]			
[0.0 0.0 0.83 0.17]				
System Cost				
[0.0 0.33 0.67 0.0]				
[0.0 0.0 0.83 0.17]	[0.05 0.11 0.77 0.07]			
[0.0 0.0 1.0 0.0]				
[0.5 0.5 0.0 0.0]		System cost	[0.27 0.31	
System Pricing			0.39 0.03]	
[0.2 0.8 0.0 0.0]				
[1.0 0.0 0.0 0.0]	[0.50 0.50 0.0 0.0]		55.04	
[0.0 1.0 0.0 0.0]			(Satisfactory)	
[1.0 0.0 0.0 0.0]				

Performance of existing systems

Figure 4.8 presents the existing performance of the UIS in KMC. It clearly indicates that the majority of the systems are operating below the satisfactory level (i.e., PI for WSS: 45.0, WWS: 41.3, SWS: 48.9, and TS: 40.1, respectively). Comparatively, the performance of the SWS is the highest and the TS performance is the lowest. The overall performance of the CIS is poor (PI of 38.2 <40.0) after the FSE operation.

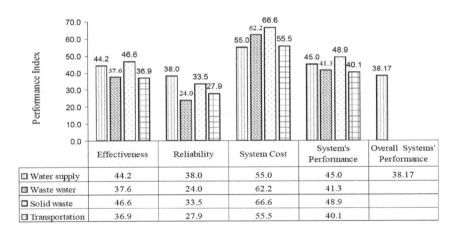

	Effectiveness	Reliability	System Cost	System's Performance	Overall Systems' Performance
☐ Water supply	44.2	38.0	55.0	45.0	38.17
☒ Waste water	37.6	24.0	62.2	41.3	
☐ Solid waste	46.6	33.5	66.6	48.9	
☐ Transportation	36.9	27.9	55.5	40.1	

Performance Parameters

Figure 4.8 Performances results of the urban infrastructure systems in KMC

More detailed analysis indicates that the effectiveness category of the WSS is satisfactory (PI = 44.2), the reliability category is poor (PI = 36.0) and the system cost category is satisfactory (PI = 55.0).

This is compared with that of the WWS, where effectiveness and reliability are poor (PI = 37.6 and 24.0, respectively), while the system cost is good (PI = 62.2). Clearly, both WSS and WWS have poor reliability and this can be attributed to irregular water supply services and frequent service interruptions, inadequate capacity of the main pipe systems, aged networks, and poor treatment capacity. On the other hand, the system costs for both WSS and WWS are good. This can be resulted due to cheaper tariffs, a better pricing structure and the high willingness to pay for the services. The results indicate that KMC must aim to improve the effectiveness and reliability category of the systems to improve the system's performance (particularly in WSS and WWS).

Sensitivity of the performance results with the assigned weights was analysed using equal weights for all of the performance indicators at each levels. The overall performance index of the systems was found to to 39.5 from the equal weight assignment, which is only 1.30 unit more that the overall performance index obtained from this analysis. One of the reasons for the insensitivity is the presence of a large number of indicators used in each dimension.

Performance-based planning

In this section, an illustration is presented on how the developed tool can be used to support investment decisions for infrastructure planning. The main objectives of the analysis, guided by KMC, aim to improve the performance of the WSS from 45.0 to 50.0 and to improve the overall system performance from its current poor level (PI = 38.2) to the satisfactory level (PI>40.0). The questions are then which components of the performance category of the WSS should be improved and by what amounts? Which system and/or systems should be improved first and by what amounts?

For the first analysis, it is assumed that the investment costs required for a unit increase in each performance category is $ 0.35 million for effectiveness, $ 0.40 million for reliability, and $0.25 million for system costs. Now, KMC plans to improve the performance of the WSS by 5.0 units, from 45.0 (i.e., to reach a higher satisfactory level). There may be several options available to achieve this objective; however, only two options have been considered in the analysis, as shown in Table 4.7. The first option increases effectiveness and reliability by 12.8 units and 8.7 units, respectively (with no change to the system cost). This option will increase the performance of WSS from 45.0 to 50.7 at a total investment cost of $7.96 million (i.e., $1.40 million/unit). The second option increases the effectiveness and reliability by 5.8 units and 8.7 units, respectively (again with no change to the system cost). The second option results in the performance increasing to 50.7, for an investment cost of $8.43 million (i.e., $1.59 million/unit). Clearly, the first option is more cost-effective and requires improvements in the service delivery, quality and system reliability categories. More specifically, future investment should be made to increase the source of supply, rehabilitation of the main pipes, limit and control of pollution sources and minimize power failures. Thus, there are several options available to improve the performance of any infrastructure system (Figure 4.9). However, the key for planning is to analyze the trade-offs between the system performance improvement and the potential costs for the investment.

Table 4.7 Performance based investment plan for the WSS

Option	System Type	Current Performance	Performance Increment	New Performance	Costs ($ Million)
1	Effectiveness	44.2	12.8	57.0	3.2
	Reliability	36.0	8.7	45.0	2.61
	System Cost	55.0	0.0	55.0	0
	Total	45	5.7	50.7	5.81
2	Effectiveness	44.2	5.8	50.0	1.45
	Reliability	36.0	16.0	52.0	4.8
	System Cost	55.0	0.0	55.0	0
	Total	45	5.3	50.3	6.25

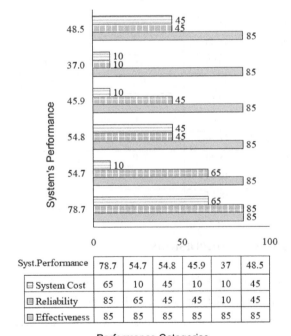

Syst.Performance	78.7	54.7	54.8	45.9	37	48.5
System Cost	65	10	45	10	10	45
Reliability	85	65	45	45	10	45
Effectiveness	85	85	85	85	85	85

Performance Categories

Figure 4.9 Relation between the performance categories and system's performance

In order to improve the performance of the UIS to a satisfactory level (PI> 40.0), two possible infrastructure investment options (options 1 and 2) have been proposed and analyzed, as shown in

Table 4.8. It is assumed that the investment costs required for a unit increase in systems performance are: $1.40 million for the WSS; $1.50 million for the WWS; $ 0.50 million for the SWS; and $2 million for the TS. The first option aims to improve the existing performance level of the WSS by 5.0 units, and the WWS by 8.7 units, while leaving the performance of the SWS and the TS unchanged. This option requires the total cost of $20.05 million, and the resulting overall system performance will be 42.8 units. The second option aims to increase the existing performance level of the WSS by 5.0 units, the WWS by 3.7 units and the TS by 4.9 units (leaving the SWS unchanged). The total cost of this option is $22.35 million, and produces a system performance of 42.7 units. Clearly, the first option gives a better return on the investment and is a more feasible option for the KMC. Nevertheless, several different options can be considered, and this tool can assist in identifying the optimal investment strategy.

Table 4.8 Performance based investment plan for the CIS

Option	System Type	Current Performance	Performance Increment	New Performance	Costs ($ Million)
1	Water Supply	45.03	5.0	50.0	7.00
	Waste Water	41.30	8.7	50.0	13.05
	Solid Waste	48.90	0.0	48.90	0.00
	Transportation	40.10	0.0	40.10	0.00
	Overall	38.17		42.79	20.05
2	Water Supply	45.03	5.0	50.0	7.00
	Waste Water	41.30	8.7	50.0	5.55
	Solid Waste	48.90	0.0	48.90	0.00
	Transportation	40.10	0.0	40.10	9.80
	Overall	38.17		42.75	22.35

4.5 Discussion and Conclusions

MCDA is the process of reaching a decision through the consideration of available alternatives, guided by many measures, rules, and standards, called criteria. Recently, there has been a growing interest in using MCDA methods in disciplines dealing with complex decision problems. This is because of diminishing available resources, growing demand, and need to incorporate multiple conflicting criteria in the decision process. The objective of undertaking MCDA in this research was to develop a MCDA aid for strategic planning and risk management of UWS. There have been many empirical studies of the applications of MCDM in decision-making process of the strategic planning of resources and risk management; however, studies on how uncertainty analysis in MCDM can deal with the potential uncertainties at each steps of decision-making process are rare.

Literature on MCDA shows that its application is growing everyday in all sectors. There are many MCDA methods already developed and applied for decision-making processes including UWS (Ananda and Herath, 2009; Hajkowicz and Higgins, 2008; Mendoza and Prabhu, 2005; Pohekar and Ramachandran 2004). The decision to be made could be a range from selecting an optimal strategy, to ranking implementation plans, to managing risks.

The general process of MCDA is defining objectives, arranging them into criteria, identifying all possible alternatives, and then measuring consequences. The literature classifies the available MCDA methods in three groups: i) *Value measurement models* commonly used either for choosing strategies or technologies, e.g., Multi-Attribute Value Theory (MAVT) (see Hostmann et al., 2005), Multi-Attribute Utility Theory (MAUT) (Keeney and Raiffa, 1993), and AHP (Saaty, 1988); ii) *Goal, aspiration and reference level models* to filter out the most unsuitable alternatives, e.g., Step Method (STEM) (Benayoun et al., 1971); and Technique for Order Preference by Similarity to Ideal Solutions (TOPSIS) (see Behzadian et al., 2012) and iii) *Outranking models* which focus "on pairwise evaluation of alternatives, identifying incomparability as well as assessing preferences and indifferences, e.g., Elimination and Choice Translating Reality (ELECTRE) (Roy and Vanderpooten, 1996) and Preference Ranking Organization Method for Enrichment and Evaluation (PROMETHEE) (Brans et al., 1986). Each of the above MCDM methods could be applied in the contexts of deterministic, stochastic, and fuzzy methods or through a combination of any of these. There are many attempts for integrating MCDA with other analytical methods as a hybrid approach.

The literature review showed that the decision results calculated from a very simple technique to a complex technique was not significantly different. Therefore, a simple approach can be equally suitable provided other operations such as selection of criteria, assignment of weights are undertaken with great care. The review also revealed that most of the MCDA are either not user friendly, not transparent or does not allow for stakeholder engagement. Thus, the need for simple and transparent MCDA methods that ensure the engagement of stakeholder was realised. Uncertainty is unavoidable in measuring and representing the performance criteria. It was well recognised that uncertainty associated with the criteria should be addressed appropriately (Mendoza and Martins, 2006). Still, most of the multi-criteria analysis methods have limitations when dealing with data sets that are uncertain, ambiguous, and linguistic (Silvert, 2000; Roy, 2005).

We have developed a new MCDA method based on the fuzzy set theory that addresses a few of these shortcomings. The proposed MCDA framework is hierarchical in multiple levels to analyse the performance of various criteria and sub-criteria at different levels under numerous dimensions such as technical, social, environmental, economic, etc. The method analyses the performance by fuzzificaition, weighting, aggregation, and defuzzificaiton called Fuzzy Synthetic Evaluation Technique (FSE). The FSE technique is one of the fuzzy set-based MCDA techniques that can handle data sets that are incomplete, ambiguous, linguistic, and uncertain. The uncertainty in the criteria is handled by representing the performance indicator by fuzzy membership functions. One of the important properties of the approach is that it allows the handling of any type of data associated with the input indicator.

The relative importance of indicators, sub-criteria, and main criteria is analysed using an AHP technique. The AHP is a widely applied technique that is very simple and easy to analyse. The weights in the performance indicators and criteria are assigned based on the ranking of stakeholders and experts working in the similar area. Despite the extensive use of AHP for the assignment of weights and addressing a variety of decision-making problems, the method has been criticized for the "rank reversal" problem. Due to the "rank reversal" problem, when a new alternative is added in an existing set of alternatives A such that it is the same with an existing alternative then the evaluations on the new set of alternatives are not consistent with evaluations on the initial set of alternatives A. Further discussion on rank reversal and other limitations of AHP applications are available in (Ho, 2008; Sipahi and Timor, 2010).

The proposed method allows for the views of stakeholders to be incorporated in two processes: i) finalising the indicators used for the assessment and their characterisation and ii) assigning the weights in AHP. In the first case, the research team presents the major indicators selected and ranges of their worst and best values. The experts and stakeholders have the role to decide and finalise the selected criteria and their quality classification (i.e., to define the range of poor, satisfactory, or good). Similarly, in the second case, stakeholders are engaged while ranking the different criteria to incorporate the local needs, and prioritize indicators and system.

The developed framework is hierarchical and can analyse the performance results as an index while allowing decision makers to diagnosis a system by focusing on: where the poor performance is located? How can it be improved? How will it contribute to improving the overall performance? Thus, the developed framework provides an interactive platform, which is required for system planning, and maximizes the performance in a cost effective way. It is flexible and equally applicable for decision-making during the risk management.

The developed MCDA framework and method was applied for the overall performance assessment of UIS in KMC. The application case shows the performance assessment of different systems in various dimensions. The case also shows how it can be applied for performance based planning while dealing with limited resources. The results show that performances of most of the infrastructure systems in

KMC are at the poor condition. The reason behind the poor performance is poor reliability and effectiveness - it means the infrastructure systems are not adequate, levels of service are poor, and social and environmental quality is very poor in general. The analysis also demonstrated how the decision maker can improve the overall performance of a single system as well as the overall system.

Risk management seeks the solutions for controlling and reducing risk in a system. This requires the selection of a potential risk strategy or an option. The risk management framework is considered as a subset of the systems performance analysis (see Figure 4.5) and follows the similar steps of analysis using different criteria in environmental, economic, social, cultural, institutional, and technical dimensions. The application of risk management has been demonstrated in Chapter 5.

It is well accepted that an integrated study helps to utilise the limited resources and improve the overall quality of life. In many cases, the causes and consequences of performance failure of interdependent systems could be beyond their own system. This is also illustrated by many real cases. For example, in Mumbai, India, during the floods that occurred in July 2005, most of the roads and buildings were more severally affected than expected due to a poor urban drainage network (canals and drains) that was blocked (Kovats and Akhtar, 2008). In the USA in August 2005, the catastrophic failure of the levee in New Orleans following Hurricane Katrina resulted in the loss of life and extensive property damage as well as severe disruption of regional and national economies (American Society of Civil Engineering (ASCE), 2005). In the UK in July 2007, around 140,000 households in Gloucestershire, Worcestershire, and the surrounding areas endured over a week without running water as a result of flooding (Severn Trent Water Limited (STWL), 2007). All of the studies concluded that the degree of loss and damage would not have been on the scale that occurred if a systems approach to infrastructure planning and management (e.g., assessment and understanding of interdependencies between system components) had been implemented. Thus, it is clear that MCDA based on a system approach of analysis offers a suitable planning and decision-making framework in this respect.

Although the developed MCDA is simple, transparent, and flexible, a few areas of further improvement are suggested. Like any other existing MCDA method, there are many distinct types of weighting methods with different philosophies each generating different weights. In many cases, the result of the MCDA is sensitive to these weights. Therefore, assigning weights to the performance indicators using subjective weighting methods is neither easy nor straightforward; it involves value judgments, subjectivity, and unfortunately, potential biases. Finding appropriate solutions to these shortcomings could be the focus of further research in this area. The credibility of performance results will depend on the types of indicators used, the quality of data available, and the weights assigned to the performance measures. We have developed a large number of indicators for the four systems.

However, further work is required to develop performance indicators that can measure the degree of interdependency between the systems.

The next Chapter presents the application of a hybrid approach of risk assessment in Mbale, Uganda, where the developed MCDA framework is applied for the risk management.

Chapter 5   Application of the Hybrid Approach of Uncertainty Analysis and MCDM Framework in UWS

The objective of this Chapter is to demonstrate the developed frameworks and methodologies for risk and uncertainty analysis and strategic decision-making in UWM. The case study considered for the analysis is in Mbale, Town Uganda. The analysis focuses on future water demand and water sources analysis under uncertainty for the year 2032. The case study analyses the heterogeneous types of uncertainty information in a single framework through uncertainty transformation and propagation methods developed in this research. The major sources of uncertainty considered in the analysis are water flow rates from the various river sources, safe yield of groundwater, leakage from the water supply system, and population growth. Both probability distribution functions and fuzzy membership functions are used to describe the uncertainty. MCS, LHS and fuzzy alpha-cut techniques are applied to propagate the uncertainty. The fuzzy set theory based multi-criteria analysis method developed from this research has been applied for the selection of an alternative water source. The analysis recommends the different strategies for alternative water sources planning to meet the future water demand. The data used for the analysis was obtained from a recent World Bank study (Eckart et al., 2012; Jacobsen et al., 2012). This Chapter is also based on the paper Khatri and Vairavamoorthy (2013c).

5.1   Introduction

Mbale is one of the most rapidly urbanizing towns in Uganda. With the high rates of urbanisation, water scarcity and wastewater management are become main challenges. The World Bank's recent study (Eckart et al., 2012) analysed the water availability in Mbale and recommended possible future water supply options for the year 2032. The report recommended an integrated urban water management scheme based on a cluster approach. In this approach, the existing municipality areas and neighbouring town areas are divided into a small-decentralised sub-area. Water and sanitation solutions for each of the clusters is analysed utilising a portfolio of local water resources including recycled grey and black water.

There is limited hydrological and river flow data to forecast the future water resources, and limited data on groundwater potential. The water distribution systems are old (more than 50 years) but National Water and Sewerage Corporation (NWSC), Uganda reports that the physical losses from the systems is approximately 12%. The population growth was forecasted assuming that the town will have a uniform population growth in all areas. Thus, the main challenges, which are common in most developing countries, are to (i) make a rational decision, despite uncertainty, for a case with limited data; (ii) handle mixed types of data (i.e., quantitative and qualitative type); and (iii) select water sources considering the multiple objectives analysis. To address those challenges, this case study aims to (i) demonstrate the developed frameworks, methodologies, and tools to address the aforementioned condition, and (ii) analyse and compare the results with the results and recommendations made by the World Bank study report.

Mbale Town

Mbale is the main administrative and commercial centre of the Mbale District and the surrounding sub-region of Uganda (Figure 5.1). It is located at 34° 10' East of the prime meridian and 1° 03' North of the Equator at the foot of Mount Elgon in Eastern Uganda. The municipality occupies an area of approximately 24.35 km². A large part of the municipality is a gentle plateau that slopes to the West. Its relief varies from 4040 ft. above sea level in the south-eastern border region to 3600 feet above sea level in the West (Ministry of Water and Environment, 2011). It lies 190 Km Northeast of Kampala resulting in advantages of access by road and railway. It has ties to the Eastern Africa region and serves as a regional link and gateway between Uganda and Kenya.

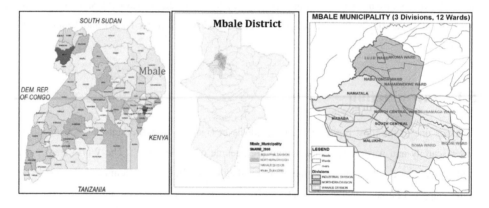

Figure 5.1 Geographic location of Mbale district and the divisions in the municipality

Mbale has a distributed rainfall ranging between 1250 mm and 1750 mm per year. The area has two peak rainy periods: March to May and October to November. Generally, the climate is warm and humid without extremes (Figures 2a & 2b).

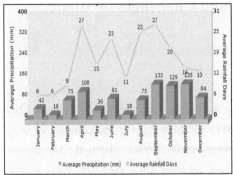 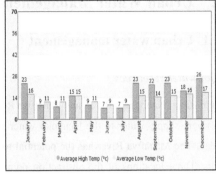

Figure 5.2 Annual variation of (a) precipitation (mm) and (b) temperature (°C) in Mbale (Source: http://www.worldweatheronline.com)

The total population of Mbale municipality in 2002 was 70,437 with an average population density of 3,354 persons per km2 and a mean household size of 4.80 persons. The annual population growth rate used by the study team is 4%, which is above the national average population growth rate of Uganda of 3.3%. The current population in the three districts of the municipality is projected based on the census results of the Uganda Bureau of Statistics (Uganda Bureau of Statistics, 2011) as shown in Table 5.1. The distribution of population within the municipality is not uniform. According to a census in 2002, the Industrial division accounted for the largest proportion of the total population with 44%, followed by the Northern division with 40%, and Wanale division with 16%.

Table 5.1 Population distribution in three district of Mbale municipality (Uganda Bureau of Statistics, 2011)

District/population	Year 2011		
	Male	Female	Total
Industrial division	20,500	20,500	41,000
Northern division	17,800	19,300	37,100
Wanale Division	6,600	7,100	13,700
Total	**44,900**	**46,900**	**91,800**

5.2   Urban Water Management and Challenges in Mbale

5.2.1 Urban water management

Major water resources

Nabijo River, Nabiyonga River, and Manafwa River are the main water sources for the town. The first two rivers supply water to the Bunkoko water works and have the potential to supply up to 5,000 m^3/day. The Manafwa River has the potential to supply about 10,000 m^3/day to the Manafwa water works. Three locations of the abstraction points and treatment works for Mbale water supply are shown in Figure 5.3. According to NWSC, surface water resources are sufficient during the wet season but the flows are low with limited abstraction potential during the dry season. For example, according to NWSC, during the dry spell it was possible to abstract a maximum of 2000 m^3/d from the Nabijo and Nabiyonga Rivers and 5000 m^3/d from Manafwa. Currently, there is no information about the ground water aquifer condition, groundwater level, potential safe yields, and water quality. Local people and other experts, however, suggest that Mbale has potential groundwater sources that require investigation.

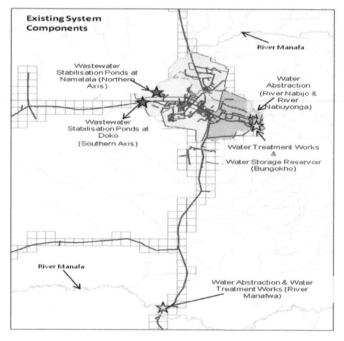

Figure 5.3 Water resources available for Mbale district

Water supply system

NWSC is responsible to supply water for the Mbale municipality. The water supply coverage in the municipality is about 71 percent with an average water consumption rate of about 70 L/c/d. About 90% of connections are domestic customers and the rest are commercial and a few industries. Most of the residents in the informal settlement get water from public standpipes. The water supply is increasingly becoming unreliable due to water scarcity. As a result, NWSC supplying the water intermittently. Recently in March 2012, water was rationed for over 12 hours a day due to low flows into the raw water dams at Nabiyonga and Nabijo. The water shortage was due to the prolonged drought that reduced river flows into the Bunkoko Water Treatment Plant.

Bunkoko water works and the Manafwa water works are the two conventional water treatment plants. Currently, the two treatment plants produce about 3800 m3/d in total, which is only a small fraction of the potentially available water source. Bunkoko water works at Mooni abstracts water from Nabijo and Nabiyonga rivers by gravity mains. The water is then collected from the rivers in a small dam located at about four km upstream from the water treatment plant. The two sources have potential to supply an amount of 5,000 m3/d, however the treatment plant design capacity is about 2,500 m3/d. The treated water is collected in a ground storage tank (10,000 m3) located next to the treatment plant. The Manafwa water treatment plant was commissioned in 1960 and is located close to the Manafwa River, which is about 17 km from the Mbale town. Water is abstracted directly from the river and the plant has a design capacity of about 9,000 m3/d but operates at a much lower capacity. The treated water is pumped to the storage tank at Bunkoko treatment plant, which then flows by gravity to the distribution system (Figure 5.3).

According to a WSSS (2012), the Mbale water supply system has a distribution system of about 270 km. About 70 percent of the pipe material is asbestos cement and cast iron. The sizes of the pipes used are 2" (more than 180 km) to 12" (about 30 km). Every day there are approximately 10-20 repairs made. According to NWSC (2012), non-revenue water (NRW) is about 10-15 percent and the physical water loss is estimated to be about 12 percent (National Water and Sewerage Corporation, 2012).

Wastewater system

The sewer coverage of Mbale is nearly 30 percent (Figure 5.4). Most of the central business district is connected to the sewer system while the rest of the population uses onsite sanitation systems - mainly pit latrines and septic tanks. The sewer pipes are made of concrete and steel (250 and 300 mm diameter). The total length of the sewer network in Mbale was about 27 km when NWSC took over and currently stands at about 31 km. The sewer system expansion in the town has been slow because of the associated high cost. There exists two lift pumping stations in the network and NWSC plans to extend the sewer network based on a centralized waste stabilization pond treatment system.

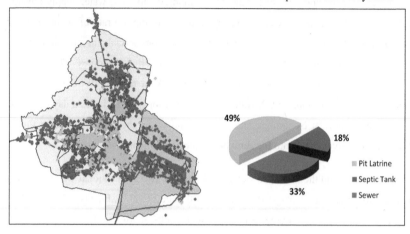

Figure 5.4 Wastewater network in Mbale town

Currently, the town has two wastewater treatment plants (waste stabilization ponds): one for the Northern axis and another one for the Southern axis areas. The two waste stabilization ponds treat about 850 m^3/d. According to NWSC, they operate at about 33 percent of their design capacity. An abandoned conventional wastewater treatment plant also exists in the town. The effluent from the stabilization ponds is discharged to the wetland next to the pond. Residents are now using the wetlands for agriculture to grow rice, maize, vegetables, etc.

The municipality manages urban drainage and solid waste. According to Mbale municipality, about 41.4 tons of waste is collected per day of which 82 percent is organic waste, 17 percent is plastic, and six percent is metal and glasses. Solid waste collection is irregular in the central part of the town or non-existent in the fringe area. Even when there are few communal rubbish dumps, these are not emptied regularly causing serious public health problems. The town does not have an engineered drainage system. Storm water is allowed to flow along the natural drainage paths (small streams and runnels) as well as along the streets.

5.2.2 Major UWM challenges

The main UWM challenges in Mbale town can be summarised as follows:

- Due to (i) increased human settlement and agricultural activities in the upstream of the supply sources and (ii) reduced flow during dry season due to climate change (though no study exits) and the amount of water available for water supply is expected to reduce in the future. However, there is no detailed study report on the river catchment and water sources that are available for the water supply in Mbale. In addition, there is no detailed information on the water quality of the rivers; there is a concern that upstream agricultural activity may be causing negative water quality impacts. It has been reported that landslides from the hills contribute to siltation at the water abstraction dams creating occasional pumping failures.

- The close proximity of the Bunkoko water works to the Mbale town is an advantage. However, the limited operational capacity of the water works means that the full potential of the water resources cannot be exploited. One of the reasons mentioned by NWSC for the limitations of operational capacity was the undersized transmission main. The dependence on Manafwa water works, which is 17 km away from the town, increases the water production costs of NWSC especially energy costs due to pumping.

- The distribution system has been growing by extending the existing system to reach the new settlements. This has resulted in pressure deficits in some locations. In addition, the municipality has development plans to build high-rise buildings, especially in the main business district and along the main roads. The pressure in the distribution system is not able to supply such high-rise buildings. NWSC has provided booster-pumping stations in two locations in the town.

- Water supply is intermittent (17 hours per day on average). The main reason for intermittency is inadequate water source due to droughts during the dry season of the year (although people think this is due to climate change impacts despite no systematic research report available to support this). This exposes the water supply system to cross contamination and could affect water quality. In many locations, the distribution pipes are crossing foul water bodies that have potential risks of cross contamination. In addition, contaminants from storm water runoff can enter the water supply in the presence of leakage points where water pipes are crossing storm drainage channels.

- The areas outside the central business district have onsite sanitation systems that consist of septic tanks and pit latrines. Although the septic tanks are well operated and maintained, many of the pit latrines are reported to be dysfunctional, and in some locations overflows during the rainy season occur. Faecal waste management is a major problem in peri-urban areas as they lack even pit latrines and septic tanks. Discussions with the stakeholders indicate that in some cases the contents of septic tanks are discharged into drains during rain events.

- The sewer pipes experience frequent blockage and breakage, especially in the cases where storm water flow is deliberately connected to sewer lines. This is the case in the central business district area where redevelopments obstructed the drainage system so that the storm water runoff is intentionally discharged into the sewer pipe. Due to these, combined effect sewage overflow during the rainy season is a serious public health issue. In addition, a higher rate of leakage in the sewers and the frequent breakdown of the existing sewer systems are believed to cause problems of cross contamination of the water supply.

- There were some reports of waterborne disease outbreaks in the town. These mainly affected water users who drew water from alternative sources (such as boreholes and springs) and not from the NWSC supplied water.

- A number of wetlands have been encroached by settlers who have blocked the storm water flow.

- Unmanaged solid waste frequently blocks the natural storm water flows leading to flooding of low laying area and disposing of the solid waste into the river.

Despite the many urban water management challenges in Mbale, water scarcity is one of the major challenges. This research focuses only on the water demand and supply analysis. As presented earlier, the objective is to demonstrate the developed framework and methodologies of uncertainty and risk analysis for a strategic decision making process in UWS planning.

5.3   Data Availability for the Analysis

The data used for the analysis was collected from the UWM study team of the World Bank report. The experts and stakeholders referred to in this case study are experts involved in data analysis for preparing the World Bank report, experts working for NWSC in Uganda, technical staff working for the Mbale municipality, and other local consultants and stakeholders who were consulted during the various phases of this research and data analysis.

5.3.1 Cluster and socioeconomic condition

The World Bank report divides the town and its emerging areas into seven Clusters for analysing the existing and projected urban growth patterns (Figure 5.5). The report presents several criterions used to select a Cluster, such as topographical features, natural drainage patterns, potential for a portfolio of water sources, planned future development, extent of existing infrastructure, etc. Clusters 1 to 5 are new Clusters, cover a total area of 1763 ha., and do not have well developed infrastructure. Therefore, water can be supplied locally from groundwater, springs, and local streams. Wastewater is managed using onsite sanitation. Cluster 6 has already developed water supply infrastructure systems. The study team suggested continuation of a centralised mode of urban water management for this Cluster.

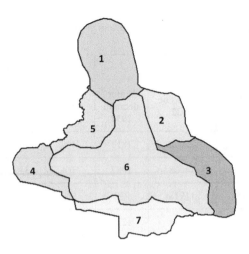

Figure 5.5 Proposed Clusters for the expected growth of Mbale

The report classifies the socio-economic condition of each Cluster into four categories based on existing housing types as presented in Table 5.2. The water consumption rate and mode of water supply connection are also differentiated according to the economic structure. Table 5.3 shows the characteristics of each Cluster in terms of population served, density and settlement types, and future water demand presented in the report. On classification, the socio economic status of Clusters' 2 and 4 population are comprised of mainly high and middle-income communities; Cluster 3 and 5 are comprised mainly of middle and low-income communities; Cluster 1 has more high-income communities located along the main road. Cluster 6 represents the existing central core of Mbale that has limited spatial growth and high resulting urban density. Cluster 7 is the forest area and it is not expected to develop.

Table 5.2 Socio-economic classes and water consumption rate

Housing categories & characteristics	Higher income group: category - 1	Medium –higher income group: category- 2	Lower medium income group: category -3	Poor income group: category- 4
Water consumption rate in LPCD	110	80	60	30
House Connection type	In-house connection with more than two bath rooms with flushing toilets, gardening	In-house connection, and bath rooms with flush toilets	Yard connection and single bath rooms with flushing toilet	Collection from public tap stand post and onsite sanitation

Table 5.3 Characteristics of the Clusters proposed for the analysis

Cluster No.	Area (ha)	Population 2012	Population 2032	Water demand 2032	Settlement types (%)			
					Cat 1	Cat 2	Cat 3	Cat 4
1	559	11348	23020	1813	10%	40%	35%	15%
2	283	6905	14008	1087	10%	35%	40%	15%
3	384	5875	11918	850	5%	35%	35%	25%
4	246	7958	16144	1272	10%	40%	35%	15%
5	291	13386	27155	2014	5%	40%	35%	20%
6	1362	62130	126037	9856	5%	50%	30%	15%
Total	3125	107602	218281	16893				

5.3.2 Future population growth and urbanization

The World Bank report has recommended a Cluster approach of urban planning where Mbale UWS is managed by a decentralised and integrated approach. The report argued that the existing central district be 'ring fenced' in terms of its infrastructure so that its centralized water supply system is not further expanded into the new emerging areas around its boundaries. The emerging areas on the boundaries should be viewed as a collection of urban Clusters that have their own autonomous decentralized water and wastewater system. In accordance with the concept of adaptation to specific contexts, the actual location, shape and size of a Cluster, the selected treatment technologies for the Cluster have to be assessed on a case-by-case basis (but is to be guided by the principle "as small as possible, as big as necessary").

The report argues that Mbale is an expanding town with an annual population growth rate of about 4%. A small proportion of this growth will take place within the boundaries of the municipality, but much of it will take place in the surrounding emerging areas. The fact that these emerging areas do not have mature infrastructure and urban planning means that there are real opportunities to implement innovative solutions for the provision of water and sanitation. The World Bank study team propose this Cluster-based development plan in these emerging areas to implement a radically different system configurations: where surface water, groundwater, and storm water are combined as potential sources; and where innovative solutions are applied that allow source separation of wastes and implementation of reclamation schemes (wastewater recycling and nutrient and energy recovery schemes). The population density and area proposed for the future growth in the World Bank report is presented in Table 5.4. As shown, the current projection for future population of new growth areas is 159,300.

Table 5.4 Extended areas of future growth from different parishes in Mbale

Perish	Population (2012)	Area (Ha)	Density	Area included in future urban growth
Namabasa	13000	1575	8	223.6
Namunsi	14200	1280	11	167.7
Nabweya	5800	573	10	169.8
Lwasso	7000	913	8	76.8
Bumboi	10400	549	19	192
Mooni	4500	333	14	115.2
Tsabanyanya	10300	1483	7	86.1
Municipality	94100	2057	46	2057
Total	**159,300**	**8,763**		**3088.20**

5.3.3 Future water demand

The World Bank report (Jacobsen et al., 2012) estimates the future water consumption in each Cluster based on the four classes of socio-economic condition as shown in Table 5.3. Future demand was estimated assuming 100 percent water supply coverage and non-revenue water of 15 percent. The total water demand forecasted for Mbale in 2032 is 16218 m3/day. The potential sources of water supply and their capacity are presented in Figure 5.6. As shown, it is based on the Cluster approach to encourage the decentralised uses of local water resources and management of the wastewater. Future water demand has been met by using the different natural water resources and recycling of grey water (see, Figure 5.6 for detail).

In the report, a list of technologies was suggested for the emerging areas of the Mbale town (except Cluster 6) that includes Decentralized Wastewater Treatment Systems (DEWATS) for greywater and blackwater, Soil Aquifer Treatment (SAT) for treatment of greywater effluent, advanced water treatment for treatment of greywater effluent, conventional water treatment for treating surface water or mixed surface water and greywater effluent, and disinfection for treatment of groundwater. For Cluster 6, waste stabilization ponds were proposed.

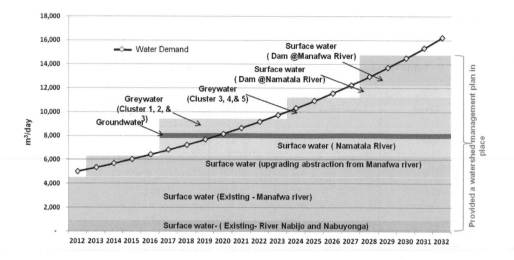

Figure 5.6 Future water demand of Mbale town and existing supplying sources (World Bank, 2012)

5.4 Methods of Analysis

Framework for analysis

The water management challenges in Mbale are analysed based on the risk assessment framework developed from this research and presented in Chapter 2 (reproduced in Figure 5.7 for clarification). The following points are considered during the analysis (Khatri and Vairavamoorthy, 2013c).

- The objectives of using the developed framework are to (i) analyse the future water demand, (ii) assess the risk of failing the future water demand from the existing resources, (iii) select of alternative sources, (iv) suggest the risk management options and (iv) compare the results with the World Bank report.

- Major uncertainties associated with the input parameters, including the rate of annual population growth, river water discharge, safe groundwater yield, and rate of water leakage from the water supply systems, are analysed. The field data is represented either by the probability distribution functions or by membership functions.

- As discussed, the uncertain information is represented either by the probability distribution functions or fuzzy membership functions (as an interval) based on the information contained by the data. The PDFs are used when a strong statistical information or high-order statistical moments is available. Experimental data needed to construct this information is often expensive and consequently no data, or only a small collection of data points, may be obtainable in the real cases. In these cases, ''expert opinion'' is used in conjunction with the available data to produce weak inferential estimates of parametric characteristics, often in the form of lower and upper bounds. The fuzzy set theory based approach has been considered as an alternative solution. The experts consulted during the analysis process are the technical staff working at NWSC Uganda and the research team of the World Bank report at the University of South Florida.

• The hybrid approach of uncertainty analysis, fuzzy set theory based risk assessment technique, and fuzzy set theory based multi-criteria analysis developed from this research have been applied. The algorithms for conversion of one form of information to the other and vice versa are as discussed in Chapter 3.The likelihood of the risk is estimated by potential water scarcity after 20 years of time in Mbale that has been calculated by a developed water demand model. The consequences are the measure of the number of people affected due to water scarcity.

• Fuzzy set theory based MCDM framework is applied by selecting the water sources for the development and risk management.

• The results are presented for each of the Clusters and overall town with uncertainty. Therefore, it will be represented by either probability distribution function or fuzzy membership function.

The next section presents the analysis and results based on the aforementioned approaches and risk assessment framework presented in Figure 5.7.

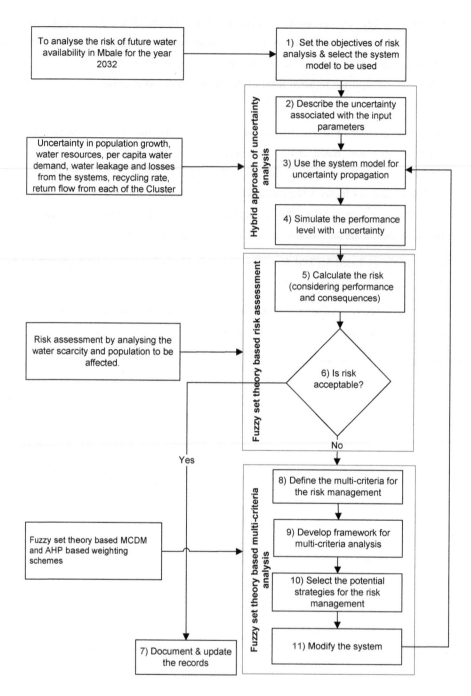

Figure 5.7 Decision making framework for the future water resources analysis in Mbale town

5.5 Analysis and Results

5.5.1 Analysis of future water demand in Mbale

Population growth rate

The population growth rate is uncertain due to various unknown factors such as actual rate of fertility, mortality and in migration in Mbale. Mbale municipal governance expects that the population growth rate in Mbale will be similar to the other smaller neighbouring towns and cities in Uganda. In this analysis, population growth rate is analysed considering a growth pattern of similar towns and cities in Uganda. Therefore, the annual growth rate considered for the analysis is from the set of [2.72, 2.93, 2.94, 2.96, 2.97, 3.31, 3.37, 3.57, 3.6, 2.69, 3.56, 3.58, 4.5, 4.5, 4.0, 4.0, 4.0] population growth that has been observed in Mbale in the past, and other smaller towns in Uganda. The sample data thus considered is simulated by 10000 MCS runs. The mean annual growth rate of population in Mbale by MCS is 3.48% with standard deviation (STD): 0.57 (Figure 5.8).

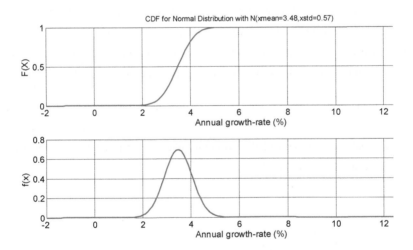

Figure 5.8 The annual population growth for Mbale after MCS

The total population forecasted for Mbale based on the geometric growth model for 2032 is shown in Figure 5.9. As shown, the forecasted population with 95% confidence interval will be [179,582, 260,052] with mean of 216,778 in year 2032.

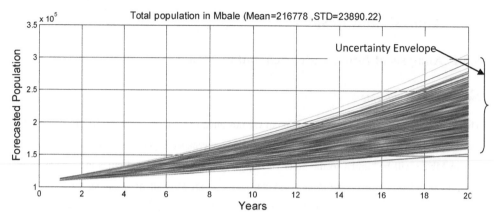

Figure 5.9 Total forecasted population in Mbale in 2032 after MCS

Population forecast for Clusters 1 to 6

The population in each Cluster was forecasted using the annual population growth rate generated and discussed in earlier sections. The mean of annual population growth rate and STD used for the analysis are 3.482% and 0.575, respectively (see Figure 5.8). The total population was forecasted assuming a geometric rate of population growth. The results of the forecasted population for Clusters 1-6 using the normal PDF and 10,000 runs of MCS are presented in Figure 5.10.

As shown in Figure 5.10, Cluster 3 contains the least population whereas Cluster 6 has the highest population. The rate of change of uncertainty envelope (i.e., difference of upper and lower level of 95% confidence interval) changes with the sizes of population. As a result, the uncertainty envelop of Cluster 6 is more than Clusters 2 and 3.

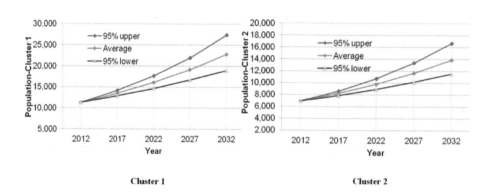

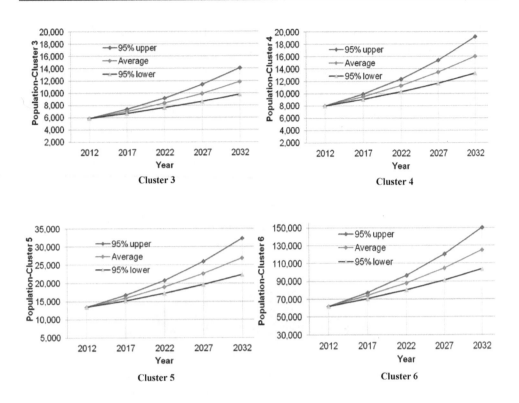

Figure 5.10 Forecasted population for Clusters 1 to 6 in Mbale

Average rate of daily per capita water demand for Clusters 1 to 6

Economic structure of each Cluster

Daily water consumption rate per capita in each cluster will depend on the socio-economic and climatic factors. Therefore, analysing the per capita water demand will require the micro component of water demand. The gross daily water consumption based on socio-economic classes used in the World Bank report is shown in Tables 5.2 and 5.3. Due to limitation of data on climate change and impact on water demand, climatic affect on water demand is not considered in this analysis.

The uncertainty associated with the percentages of certain economic classes (e.g., higher income, medium-higher income population) in each cluster is represented by the uniform PDFs. Therefore, the daily water consumption rate, rate of recycling potential from each of the clusters, and total population of a particular economic class are represented by PDFs. For example, for Cluster 1, the percentage of high income people will be [5, 15] with water consumption rate of [110,120] LPCD; percentage of high-medium income people will be [30, 40] with water consumption rate of [60,100] LPCD; percentage of low-medium income people will be [30, 50] with water consumption rate of

[50,70] LPCD; and percentage of low income people will be [10, 20] with water consumption rate of [20,40] LPCD. It is noted that the minimum and maximum range are estimated after consultation with the local experts and according to the neighbouring areas of the clusters and potential future growth. After 1000 LHS run, the resulting PDF for the daily average per capita water demand for Cluster 1 with 95% confidence interval will be [74.61 60.38 90.23] (Figure 5.11).

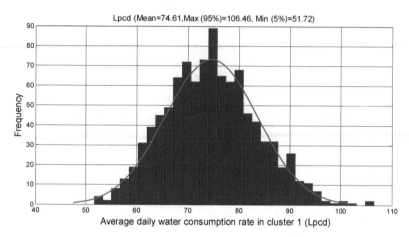

Figure 5.11 Average daily water consumption rate in Cluster1in litres per capita per day

The average rate of daily per capita water demand in each cluster is calculated by considering the different economic levels of the people living in each cluster and the rate of water consumption. However, the percentage of population living in each cluster and rate of water consummation are very uncertain and thus represented by uniform PDFs. The data used for the analysis for Cluster 1 is shown in Table 5.5. The resulting average daily water consumption rate for clusters after 1000 LHS is shown in Figure 5.12.

Figure 5.12 shows that the ranges of uncertainty in LPCD estimation are similar for each cluster. The daily water consumption rate per capita for Clusters 2 and 4 are slightly higher compared with the other clusters because of the higher percentage of richer people living in those clusters. Cluster 6 has mixed settlement types resulting in an average daily per capita water consumption lower than Clusters 2 and 4.

Table 5.5 Socio-economic classes and water consumption rate in Cluster 1

	Higher income group	Medium–higher income group	Lower-medium income group	Poor income group
Water consumption rate in LPCD	110-120	60-100	50-70	20-40
Percentage of population	5-15	30-40	30-50	10-20

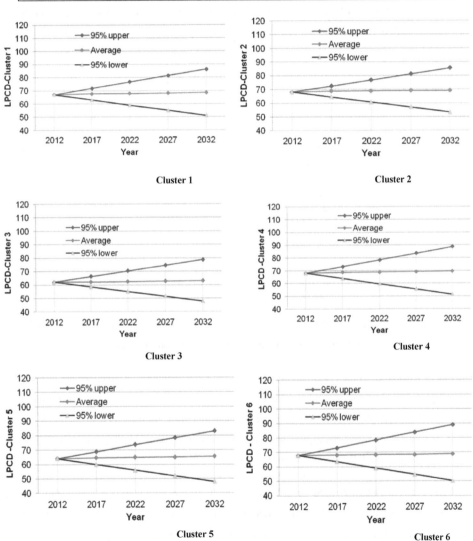

Figure 5.12 Forecasted LPCD for Clusters 1 to 6 in Mbale

Leakage rate from the water supply system for Clusters 1 to 6

According to NWSC, current rate of water leakage from the water supply system is about 12-15 percent. This rate is relatively low compared to the other similar towns in Uganda. It is expected that water supply quantity will be increased in the future to meet the future water demand. It will result in an increase in the size of the networks and water supply pressures. As a result, maintaining the water losses in the current rate will be challenging for NWSC. There are no direct techniques for predicting the future water loses. However, after consultations with the experts working for NWSC and other experts involved in the World Bank study, the rate of water leakage from the water supply system has been represented by trapezoidal shaped fuzzy membership functions. The likely values of fuzzy membership functions are 15 to 20 percent with minimum and maximum ranges of 10 percent and 25 percent respectively (Figure 5.13). Future water demand analysis in a cluster level as well as town scale will be based on this leakage rate.

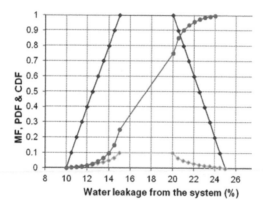

Figure 5.13 Total water leakage rate from the system in a fuzzy form and Transformation of fuzzy information into the probability distribution function

The rate of leakage from the systems represented by the fuzzy membership functions are transformed into the probabilistic form by applying the evidence theory based algorithms for transformation as presented in the Chapter 3. The resulting transformed functions are presented in Figure 5.13. As shown, top line is the membership function, the lowest line represents the PDF after transformations and the middle line is CDF generated from the PDF.

The CDF, thus generated was used for LHS. After 1000 LHS run, total water leakage quantity in each Cluster was calculated. Daily water leakage rate from other Clusters were calculated by employing the same principles and techniques and results are presented in Figure 5.14.

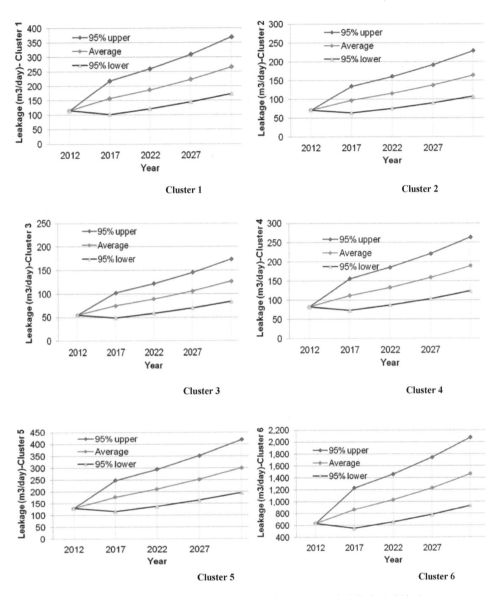

Figure 5.14 Forecasted leakage rate for Clusters 1 to 6 in Mbale (m3/day)

The total water demand for Clusters 1 to 6

The total water demand in each of the Cluster is forecasted by considering the population, their per capita water consumption rate, and leakage and losses from the system. The results of total water demand forecasted after 1000 simulations of LHS are presented for other Clusters are shown in Figure 5.15.

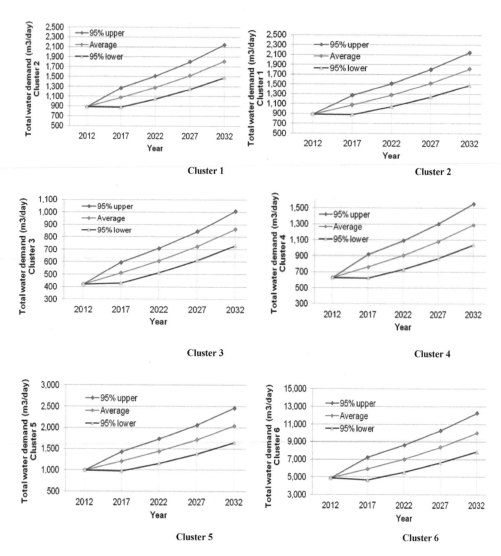

Figure 5.15 Forecasted total water demand (m3/day) for Clusters 1 to 6 in Mbale

5.5.2 Analysis of available natural and recycled water sources

Surface water sources

Nabijo River, Nabiyonga River, and Manafwa River are the main surface water sources for the town (see Figure 5.3). As presented earlier, the first two rivers supply water to the Bunkoko water works and have the potential to supply up to 5,000 m³/d. The Manafwa River has the potential to supply about 10,000 m³/d to the Manafwa water works. According to NWSC, Namatala River can be an alternative source of water to Mbale that requires a dam constructing near the town. The potential supply from Namatala will be approximately 2,000 to 5,000 m3/d. However, there are not any study report and observation on annual variation of river flow. This is the data limited situation with a higher degree of uncertainty.

In order to develop a synthetic probability distribution for each of the river, it is assumed that annual river flow will follow the same trends of rainfall patterns as shown earlier in Figure 5.2. The resulting river flow distribution derived for each river is shown on left side of Figure 5.16. For example, in Namatala River, average flow rate varies from 2,000 to 5,000 m3/day. The best fit of the generated data shows the approximate normal distribution for each of river. Based on this information, resulting PDFs generated for each river after 1000 simulations of LHS run is shown in right side of the Figure 5.16. The generated PDFs of river flow are considered for the analysis.

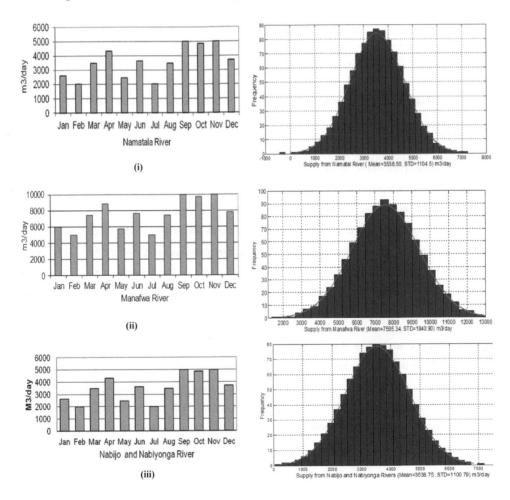

Figure 5.16 Synthetic daily distribution of three rivers water flow (m3/day)

The total water resources available from all the rivers sources are forecasted by combining three sources by 1000 LHS simulations. The resulting water quantity with 95% of confidence interval is shown in Figure 5.17.

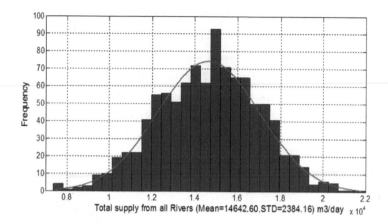

Figure 5.17 Total water supply potential from all rivers in Mbale (m3/day)

Groundwater sources

According to the World Bank report (2012), there are no any field reports and records available about the groundwater resources. However, the report has proposed a groundwater source in Cluster 1 with a safe yield of 48.96 m3/day. The number and safe yield of the groundwater sources is very uncertain to predict without any supporting information. This research uses a trapezoidal shaped membership function to represent the uncertain groundwater source. The shape and safe yield of the well is shown in Figure 5.18a. As shown, the likely discharge will be 35-40 m3/day with minimum and maximum of 25 m3/day and 50 m3day. It is noted that the shape of the membership function was generated after consultation with the local people and NWSC technical staff working in Mbale.

The information contained by the fuzzy membership function is transformed into the probabilistic form by applying the fuzzy to probability transformation algorithms as presented in the Chapter 3. The resulting transformed functions are presented in Figure 5.18b. As shown, top line is the membership function, the lowest line represents transformed PDF and the middle lines is CDF generated from the PDF. The total groundwater supply forecasted using this CDF and after 1000 LHS is shown in Figure 5.19.

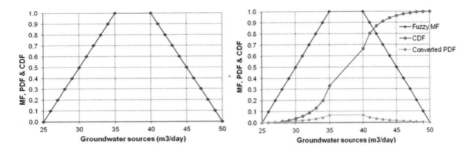

Figure 5.18 (a) Proposed fuzzy membership function for groundwater yield, (b) transformation of fuzzy information (safe groundwater yield in m3/day) to probability distribution function.

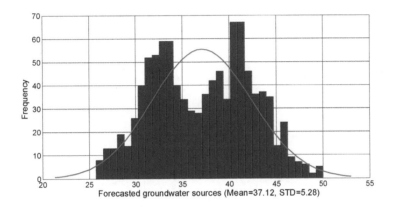

Figure 5.19 Total potential groundwater source for Cluster 1 in Mbale (m3/day)

The total water sources available in Mbale were combined by 1000 LHS simulation. The result of simulation for total water supply sources with 95% confidence interval is shown in Figure 5.20.

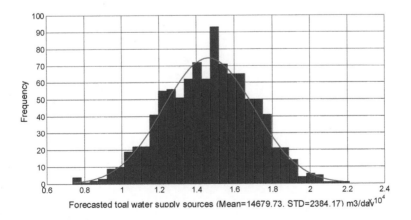

Figure 5.20 Total natural water supply potential in Mbale (m3/day)

The Recycle water

Greywater reuse in each Cluster 1 to 6

The recycling rate in each cluster is estimated by considering the daily water consumption rate of the different economic classes. After consultation with experts working for NWSC, it is assumed that the recycling rate will be possible only from the high-income, medium-high, and medium-income population group. For example in Cluster 1, the proportion of recycling rate from the high-income and medium-high income population will be about 65 percent with STD of 15 percent. It will result by assuming approximate micro water demand by 20 percent for showers, 30 percent for laundry, and 15 percent for kitchen use, respectively. Similarly, the recycling rate from the medium-income group will be about 35 percent with STD of 15 percent. The resulting recycling rate of grey water after 1000 times LHS is shown in Figure 5.21 (1). The same process was applied for the rest of the other clusters. Results are shown in Figure 5.21.

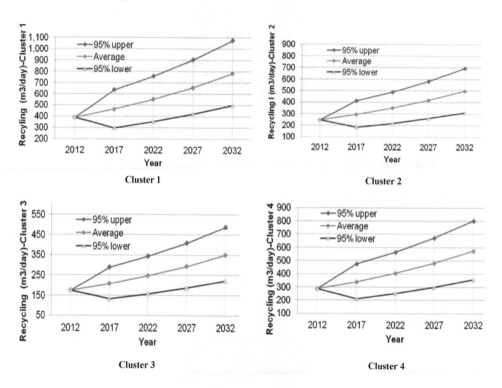

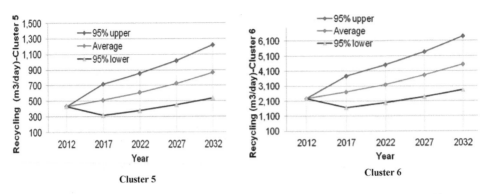

Figure 5.21 Forecasted Greywater recycling potential for Clusters 1 to 6 in Mbale (m3/day)

Cluster 6 is the existing main municipality area. The town centres, administrative units, and industrial and business units are in this cluster. Therefore, it may be questionable if the recycled water from Cluster 6 can be collected or not. Thus, total recyclable water in Mbale is analysed including and excluding Cluster 6, and the results are presented in Figures 5.22 (i) and (ii).

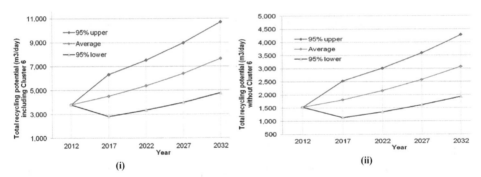

Figure 5.22 Total Greywater recycling potential (i) including and (ii) excluding Cluster 6 (m3/day)

Recycling rate of Blackwater in Cluster 1 to 6

Similar to recycling rate estimation, the Blackwater is expected only from the high-income, medium-high and medium income population group (Table 5.2). The quantity of Blackwater after simulation of 1000 LHS sample for Cluster 1 is shown in Figure 5.23 (1). The results of estimated Blackwater for other Clusters are presented in Figure 5.23.

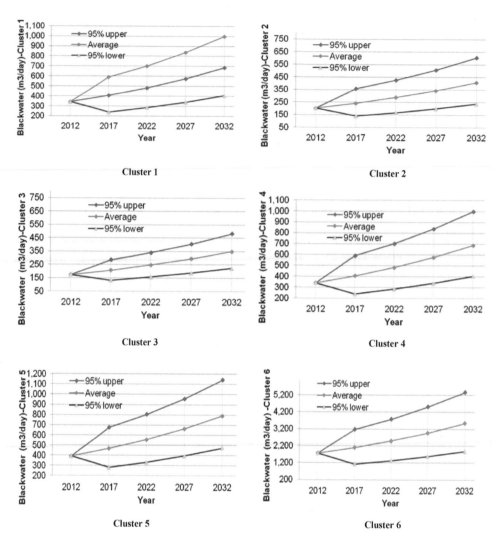

Figure 5.23 Forecasted Blackwater recycling potential for Clusters 1 to 6 in Mbale (m3/day)

5.5.3 Risk assessment for water scarcity

Risk assessment

Water demand and supply situation in Mbale town is analysed by comparing the total demand required and total water resources available. The total water demand and water resources situation after combining all the water sources is shown in Figures 5.24 (a) and (b). The lower envelop is the total water demand with 95% confidence interval [13,667, 20,777] m3/day, whereas upper lines represents the total supply situation with 95% confidence interval [10,665, 18,479] m3/day after combining all the natural water sources.

170

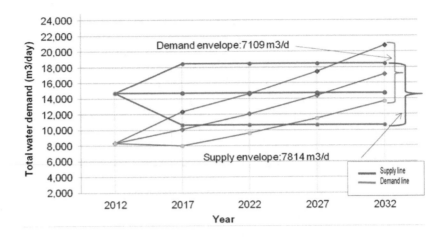

Figure 5.24 (a) Total water demand and supply situation analysis in Mbale up to year 2032
(m3/day)

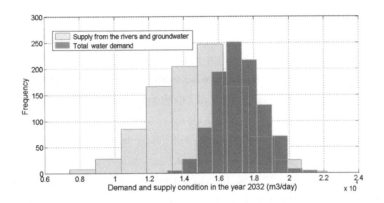

Figure 5.24 (b) Total water demand and supply situation analysis in Mbale in year 2032
(m3/day)

Figure 5.24 (b) shows that existing water sources will be unable to meet the future demand of Mbale. The demand and supply analysis shows that there a water deficit of 2,435 m3/day, approximately 14.20 percent water demand for average water demand 17,115 m3/day compared with the total natural water supply of 14,680 m3/day.From the uncertainty envelop of supply and demand, the deficit ranges show [11.06, 14.22, 21.96] in percentages. Therefore, the future water demand have to be met by available alternative water sources, such as building a dam, or reusing recycled water, or rain water harvesting, or other water demand management options. The potential future water demand management options could be controlling leakage and losses from the systems, progressive pricing, or use of water saving devices. Figure 5.25 represents the likelihood of risk (red dotted line) based on this deficit range.

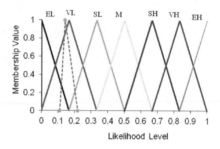

Figure 5.25 Likelihood of risk of water availability in Mbale

The likelihood of risk after fuzzification with max-operation will result in the membership values of [0.32, 0.92, 0.21, 0, 0, 0, 0], as shown in Figure 5.25. The normalised likelihood of risk will be [0.22, 0.63, 0.15, 0, 0, 0, 0]. As shown in Figure 5.30, the likelihood values are in extremely low to very low ranges.

The severity components of the risk Equation represents the number of people affected by water deficit (Figure 5.26). The future water demand water in Mbale for year 2032 will be [13,667, 17,115, 20,777] m3/day. If the existing water sources of 14,680 m3/day are considered, the average water deficit in year 2032 will be 2,435 m3/day. Given the future daily water consumption rate of 70 LPCD, the deficit will affect approximately 34,788 people daily. Thus, the severity values with respect to the affected percentage of the population to the total population within 95% confidence interval [179,581, 260,052, 216,777] will result in 13.37%, 16.04% and 19.37%, respectively. This will result in the normalised severity values of [0.13, 0.16, 0.19] respectively as presented in Figure 5.26 and shown by dotted red lines.

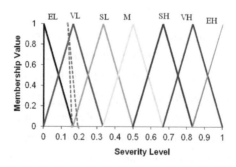

Figure 5.26 Level of severity from the failure of supply (unmet future demand) from the existing water sources

172

The magnitude of risk of water availability in Mbale in 2032 is quantified using the risk Equations shown in Appendix III. The magnitude of the risk is estimated by fuzzy alpha-cut technique as shown in Figure 5.27 (b). Both likelihood of risk (Figure 5.25) and severity (Figure 5.26) are very low and thus the resulting risk will be very low. The calculated fuzzy results of each alpha-cut are shown in Figure 5.27(b). As shown, the risk will be product of likelihood and severity of risk. The fuzzy alpha-cut technique has been applied for the calculation.

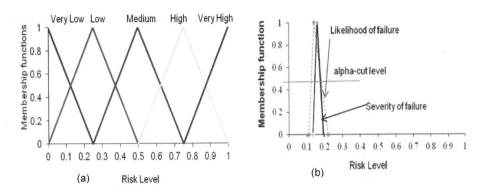

Figure 5.27 (a) Risk classification, and (b) risk calculation by fuzzy alpha-cut technique

Figure 5.27(a) shows the risk classification used for this research and Figure 5.27(b) shows both the likelihood of risk and consequences. As per our risk classification, the resulting risk lies within the range of very low. However, Mbale town must ensure water for the entire population with the desired level of services. Therefore, deficit water must be supplied or managed by alternative water sources. In order to manage the risk of water deficit, the World Bank report suggested the recycling and reuse of water in addition to augmentation by natural river sources. The next section presents the application of MCDM framework for risk management and water resources selection in Mbale.

5.5.4 Multi-criteria based risk management

A potential option for managing the risk of water scarcity in Mbale would be to supply the additional water resources to meet the demand. The current rate of water leakage in the Mbale water supply system is about 12 percent, low compared to other towns and cities in Africa. According to NWSC, the economical rate of leakage for Kampala, Uganda is about 22 percent. The lower rate of leakage and slightly higher rate of economic rate means there is less opportunity of saving water from leakage control. Similarly, the average daily per capita water consumption rate in Mbale is about 70 LPCD, which is also not higher compared to other developed towns and cities in Africa, and populations with higher rate of consumption are very less (i.e., mostly around 5 to 10 percent in a cluster). Therefore, there are much fewer opportunities to meet the exiting water demand by water demand management.

The potential alternative option of water supply augmentation could be achieved by using all the river sources, recycling, and reusing the water. The World Bank report (2012) suggested for combined action of using both groundwater, river water, and recycling water without any decision-making framework.

MCDM framework, algorithms and application

This research applies fuzzy set theory based MCDM framework for prioritising the alternative sources. The framework developed for the option analysis and algorithms developed for modelling the MCDA is shown in Figures 5.28 (a) and (b). The framework is fuzzy-based and allows the capture of most of the uncertain information. As discussed in Chapter 4, the following major steps will be required for the MCDM:

i)   Identification and classification of performance indicators;

ii) Fuzzification of the performance indicators;

iii) Weight calculation and weight assignment;

iv) Aggregation of performance indicators; and

v)   Defuzzification of the aggregated indices to produce the overall systems performance.

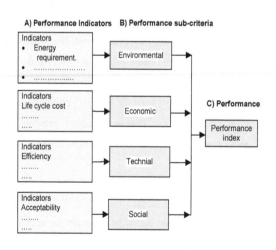

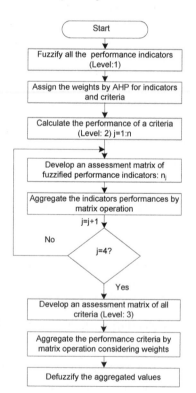

Figure 5.28 (a) MCDA framework for risk management and (b) Algorithms used for modelling the MCDA to select an option for the development

The indicators used for the analysis and their potential ranges of quality levels are presented in Table 5.6. It is noted that the quality levels (i.e., Very good, Good, Satisfactory, and Poor) are classified based on the best practices available in literature, consultations with experts working for NWSC Mbale, and the expert involved in the World Bank work of Mbale. The membership functions used for the analysis are the same as presented and discussed in Chapter 4. The triangular and trapezoidal membership functions are used for the analysis (Figure 5.29).

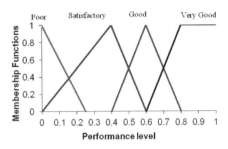

Figure 5.29 Fuzzy membership functions used to represent the performance criteria used for MCDA

Table 5.6 Indicators used for assessing the alternative option of water resources and their classification in four membership functions

Criteria and indicators	Ranges of membership functions			
	Very good	Good	Satisfactory	Poor
A. Environmental Criteria				
i) Energy requirement to lift the water from the selected sources to the system considering the head in meter	<10 m	10-20 m	20-50 m	>50 m
ii) Impact on environment with respect to the environmental flow available of total flow	<25 %,	(20-50) %,	(50-70) %,	>70 %
iii) Potential pollution load considering the total organic carbon (TOC)	<2 mg/L	(2-4) mg/L	(4-8) mg/L	> 8 TOC mg/L
iv) Nutrients recovery potential from the proposed source	>75 %	(50-75) %	(25-50) %	<25 %
B. Economic Criteria				
i) Lifecycle - unit cost considering the water supply & wastewater system	<0.25 $/m3	(0.25 -0.50) $/m3	(0.50 -0.75) $/m3	>0.75 $/m3
ii) Willingness to pay (if water is abstracted from the different sources)	Very high	High	Satisfactory	Poor

C. Technical Criteria				
i) Additional space requirements (land requirements to build the system)	<25 %	(25-50) %	(50-75) %	>75 %
ii) Reliability of source (probability of not to fail the supply)	>90	(75-90) %	(60-75) %	<60 %
iii) Efficiency of water uses (quantity to serve by the same)	>90	(75-90) %	(60-75) %	<60 %
iv) Technology requirement (starting from simple to modern/advanced types requirement)	Very simple	Medium	Modern	Very advanced
D. Social Criteria				
i) Acceptability of services /technology (based on the cultural values)	Very high	High	Low	Very low
ii) Risk to public health (potential impact on public health if the technology fails)	Very high	High	Low	Very low

The alternative options available for augmenting the water sources are: (i) water collection from Manafwa river by dam construction, (ii) water collection from Namatala river by dam construction, and (iii) Greywater recycling. The field observations for each of the alternative sources are presented in Table 5.7. As shown, some of the values are in crisp form and other are in probabilistic and fuzzy forms. The values reflect the view of local stakeholders of Mbale and experts working in NWSS.

Table 5.7 Indicators used for assessing the alternative option of water resources in Mbale

Observed and assigned values	Manafwa dam	Namatala dam	Greywater recycling
A. Environmental Criteria			
i) Total head required to lift the water from the selected sources to the system (to asses the energy requirement in a system)	56 m	21 m	Less than 10 m
ii) Impact on environment with respect to the environmental flow available of total flow	About 50%	About 40%	No impact
iii) Potential pollution load considering the total organic carbon (TOC)	2-3 mg/l	4-5 mg/l	>8 mg/l
iv) Nutrients recovery potential from the proposed source	Nil	Nil	More than 75%
B. Economic Criteria			

		0.55 to 0.65 $/m3	0.65-0.75 $/m3	0.90-1.20 $/m3
i)	Lifecycle - unit cost considering the water supply			
ii)	Willingness to pay (if water is abstracted from the different sources)	Very high	High	Poor
C.	**Technical Criteria**			
i)	Additional space requirements (land requirements to build the system)	Very low about 10%	Medium about 30%	very high - more than 75%
ii)	Reliability of source (probability of not to fail the supply)	About 80 to 85%	About 80 to 85%	About 50%
iii)	Efficiency of water uses (considering the use of recycle water , i.e., quantity to serve by the same)	< 20%	About 40%	About 90%
iv)	Technology requirement (starting from simple to modern/advanced types requirement)	Simple	Intermediate	Modern and advanced type
D.	**Social Criteria**			
i)	Acceptability of services /technology (based on the cultural values)	Very high	High	Low
ii)	Risk to public health (potential impact on public health if the technology fails)	Very low	Very low	Very high

The indicators listed in Table 5.7 are fuzzified using the fuzzy operations discussed earlier in Chapter 4 and presented in Annex II. The relative importance of each of the indicators and criteria are decided after consultation with local stakeholders in Mbale and capturing their priorities using AHP based weighting techniques as discussed in Chapter 4 (see also Equations 4.2,4.3, and 4.4 Table 4.1). A sample of weight calculation by AHP application is shown in Table 5.8.

Table 5.8 Judgement matrix of environmental criteria for Mbale

A. Environmental Criteria	**Energy**	**Impact**	**Pollution**	**Nutrients**	**Weight**
i) Total head required to lift the water from the selected sources to the system (to assess the energy requirement in a system)	1.00	1.00	1.00	3.00	0.30
ii) Impact on environment with respect to the environmental flow available	1.00	1.00	2.00	3.00	0.35
iii) Potential pollution load considering	1.00	0.50	1.00	3.00	0.25

the total organic carbon (TOC)					
iv) Nutrients recovery potential from the proposed source	0.33	0.33	0.33	1.00	0.10
Column Sum	**3.33**	**2.83**	**4.33**	**10**	**1.00**

As shown in Table 5.8, the judgement matrix consist of four dimensions of pair-wise comparison, $n=4$; Eigen value of the judgment matrix, $\lambda_{max} = 4.07$; and Saaty's average random consistency index value from Table 4.1(Chapter 4), RCI= 0.90. The consistency ratio of judgement matrix for this combination of values will be 2.97, less than 10%. Therefore, the judgement matrix is consistent. The same principles and procedures were applied for weight calculations. The results of the calculation are presented in Table 5.9.

The performances of different options are analysed using the Fuzzy Synthetic Evaluation Technique as presented in Chapter 4. The algorithm used for aggregation at different levels is shown in Figure 5.28(b). Equations 4.5, 4.6, and 4.7 were applied for the aggregation process, whereas Equation 4.9 was applied for the defuzzification. The analysis results of fuzzification, aggregation, and defuzzification for three options are presented in Table 5.10.

Table 5.9 Judgement matrix for criteria and indicators used for the analysis

Criteria and indicators	Weight
A. Environmental Criteria	**0.15**
i) Total head required to lift the water from the selected sources to the system (to assess the energy requirement in a system)	0.30
ii) Impact on environment with respect to the environmental flow available	0.35
iii) Potential pollution load considering the total organic carbon (TOC)	0.25
iv) Nutrients recovery potential from the proposed source	0.10
B. Economic Criteria	**0.33**
i) Lifecycle - unit cost for the water supply & wastewater system	0.75
ii) Willingness to pay (if water is abstracted from the different sources)	0.25
C. Technical Criteria	**0.34**
i) Additional space requirements (land requirements to build the system)	0.13
ii) Reliability of source (considering probability of not to fail the supply)	0.53

iii) Efficiency of water uses (quantity served by the unit abstraction)	0.24
iv) Technology requirement (starting from simple to modern/advanced types requirement)	0.10
D. Social Criteria	**0.18**
i) Acceptability of services /technology (based on the cultural values)	0.40
ii) Risk to public health (potential impact on public health if the technology fails)	0.60

Table 5.10 shows results in both fuzzy and index forms. For example, from the environmental perspective, the fuzzy result of Manafwa dam is [0.34 0.50 0.15 0.00] and the index value is 30.42. This means, if the Manafwa dam is built for water supply, then its environmental performance will be 34% poor, about 50% satisfactory, about 15% good, and 0% very good. Nevertheless, after the fuzzy operation, the index value will be 30.42, in the satisfactory range of the classification. Thus, the calculated results show the performance of each option under each criterion and relative performances with other options. In the same way, the performance of recycling from an environmental perspective is highest compared with two other options (i.e., 60.65). The final index value after aggregation of all the criteria is shown in the last row of the Table. From the result, it is clear that building a dam on the Manafawa River will be the most suitable and preferred option, and other second and third options will be building the dam on the Namatala River and greywater recycling. Thus, the priority for developing the systems should be based on the results.

Table 5.10 Calculated performance values of different criteria after aggregation at different levels

Calculated performance levels	Manafwa dam	Namatala dam	Greywater recycling
A. Environmental criteria	[0.34 0.50 0.16 0.00]	[0.02 0.78 0.20 0.00]	[0.19 0.06 0.30 0.45]
	Satisfactory 30.42	Satisfactory 38.00	Good 60.65
B. Economic criteria	[0.15 0.60 0.25 0.00]	[0.60 0.15 0.12 0.13]	[0.94 0.06 0.00 0.00]
	Satisfactory 36.75	Satisfactory 30.88	Poor 14.25
C. Technical criteria	[0.26 0.04 0.30 0.40]	[0.13 0.21 0.46 0.20]	[0.22 0.44 0.03 0.31]
	Satisfactory to good: 56.40	Satisfactory to good: 53.29	Satisfactory to good 45.69
D. Social criteria	[0.00 0.00 0.25 0.75]	[0.00 0.00 0.28 0.72]	[0.83 0.17 0.00

			0.00]
	Good to very good 78.62	Good to very good 77.87	Poor 16.40
E. Total performance index	Satisfactory to good 50.02	Satisfactory to good 48.02	Satisfactory 32.30

Strategy for Managing the risk of water scarcity in Mbale

Earlier analysis results showed that there is a risk of not meeting the future water demand in Mbale from the exiting sources. The public preferences for the alternative sources have been selected according to the MCDA result as presented in Table 5.10. However, this analysis does not cover the detailed cost analysis for the dam construction on the rivers to meet the additional water demand. Therefore, the preferences of the sources may vary with the change in the unit costs of water supply.

This analysis presented the uncertainty envelope within 95% confidence interval for water supply and demand (Figure 5.24b). Considering the resulting water supply gaps, the future water supply sources are analysed for the two possible options: (i) supplying by all the natural water resources and supplemented by building the dams, and (ii) supplying by the existing surface water, ground water, and supplemented by recycled greywater. For both options, three scenarios are developed considering the 95% confidence interval of the analysis, which will be more realistic than scenarios developed by simple assumptions.

Conventional approach: supplying by natural water sources

Scenario I: Water demand increases along the maximum rate of forecasting

For this scenario, the strategy for water supply will be as presented in Table 5.11 and Figure 5.30. As shown, the exiting water sources will be augmented by ground water before 2022 and then by other surface water sources. Total quantity for the augmentation required to supply from the river sources by building the dams are shown in the deficit column of Table 5.11.

Table 5.11 Demand growth in maximum rate and supply met by the rivers and groundwater

Year	Demand	Supply (m3/day)				Remarks
	(m3/day)	Maximum	Surplus	Deficit	Add	
2012	8343.07	11104.00	2760.93	-	-	Existing supplying rivers
2017	12322.42	14642.50	2320.08	-	-	Groundwater:37.20

2022	14616.21	14642.50	26.29	-	Ground water	m3/day
2027	17419.21	14679.70	-	2739.51	Dam at Manawafa	
2032	20777.00	14679.70	-	6097.30	Dam at Namatala	

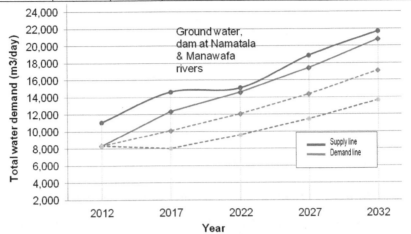

Figure 5.30 Demand and supply balance for maximum rate of water demand increase

Scenario II: Water demand increases along the average rate of forecasting

The exiting water sources will be augmented by ground water after 2022 and then other surface water sources by building the dams by 2027. The possible strategies for water supply for this scenario are presented in Table 5.12 and Figure 5.31.

Table 5.12 Demand growth in average rate and supply met by rivers and groundwater

Year	Demand	Supply (m3/day)				Remarks
	(m3/day)	Average	Surplus	Deficit	Add	
2012	8343.06	11104.00	2760.93	-	-	Existing supplying rivers
2017	10109.99	14642.50	4532.51	-	-	
2022	12040.41	14642.50	2602.09	-	-	
2027	14349.24	14679.70	330.46		Build dam at Manawafa River	Groundwater:37.20 m3/day
2032	17115.19	14679.70		2435.49	Build dam at Namatala River	

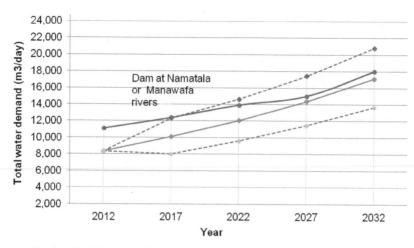

Figure 5.31 Demand and supply balance for average rate of water demand increase

Scenario III: Water demand increases along the minimum rate of forecasting

This scenario will be less likely in Mbale. If this scenario materialised, the exiting water sources will be augmented by Manawafa river source before 2022 to meet the expected demand for the future Further supply and demand analysis for this scenario are as presented in Table 5.13 and Figure 5.32.

Table 5.13 Demand growth in a minimum rate and supply met by rivers

Year	Demand (m3/day)	Supply (m3/day)				Remarks
		Average	Surplus	Deficit	Add	Existing supplying rivers
2012	8343.07	11104.00	2760.68	-	-	
2017	7997.17	11104.00	3106.58	-	-	
2022	9614.36	11104.00	1489.39	-	-	
2027	11458.48	14641.75	3183.27	-		Add Manawafa River
2032	13667.10	14642.50	975.40		-	

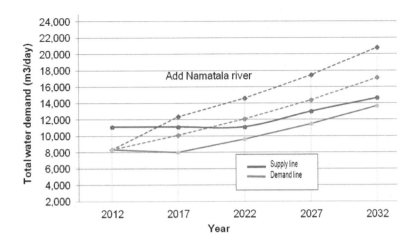

Figure 5.32 Demand and supply balance for minimum rate of water demand increase

IUWM approach: supplying by natural water sources and recycling water

Scenario I: Water demand increases along the maximum rate of forecasting

The strategies for water supply for this scenario are presented in Table 5.14 and Figure 5.33. As shown, the exiting water sources will be augmented before 2022 by Manafwa river water, groundwater, and greywater source. The additional quantity of water to be supplied from the sources is shown in the deficit column of Table 5.14.

Table 5.14 Demand growth in a maximum rate and supply met by river, groundwater and greywater

Year	Demand (m3/day)	Supply(m3/day)				Remarks
		Average	Surplus	Deficit	Add	
2012	8343.07	11104.00	2760.93			Existing supplying rivers
2017	12322.42	14642.50	2320.08			Groundwater:37.20 m3/day
2022	14616.21	14679.70	63.49		Build dam at Manafwa river	
2027	17419.21	21075.24	3656.03			Greywater: 6395.54 m3/day
2032	20777.00	27470.78	6693.78			

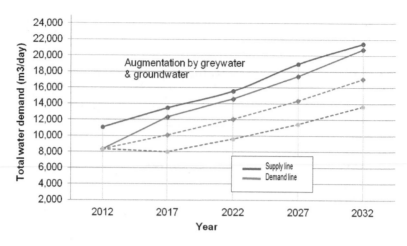

Figure 5.33 Demand and supply balance for maximum rate of water demand increase

Scenario II: Water demand increases along the average rate of forecasting

The exiting water sources will be augmented by greywater only before 2022. The demand and supply analysis with possible strategies for this scenario are presented in Table 5.15 and Figure 5.34.

Table 5.15 Demand growth in an average rate and supply met by groundwater and greywater

Year	Demand (m3/day)	Supply (m3/day)				Remarks
		Average	Surplus	Deficit	Add	
2012	8343.07	14642.50	6299.43	-	-	Existing supplying rivers
2017	10109.99	14642.50	4532.51	-	-	
2022	12040.41	14642.50	2602.09	-	-	Groundwater: 37.20 m3/day
2027	14349.24	17253.64	2904.40	-	37.20	
2032	17115.19	19827.58	2712.39		2573.94	Greywater: 2573.94 m3/day

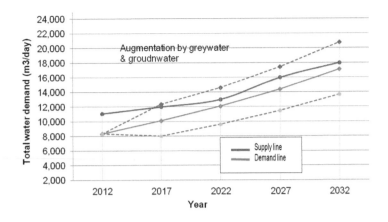

Figure 5.34 Demand and supply balance for average rate of water demand increase and augmentation by greywater supply

Scenario III: Water demand increases along the minimum rate of forecasting

For this scenario, the strategy for water supply will be as presented in Table 5.16 and Figure 5.35 As shown, the exiting water sources should be augmented before 2022 by groundwater and Greywater.

Table 5.16 Demand growth in a minimum rate and supply met by groundwater and greywater

Year	Demand (m3/day)	Supply (m3/day)				Remarks
		Average	Surplus	Deficit	Add	
2012	8343.07	11103.75	2760.68	-	-	Existing supplying rivers
2017	7997.17	11103.75	3106.58	-	-	
2022	9614.36	11103.75	1489.39	-	37.20	Groundwater: 37.20 m3/day
2027	11458.48	13713.95	2255.47	-	2573.94	Greywater: 2573.94 m3/day
2032	13667.10	13713.95	46.85			

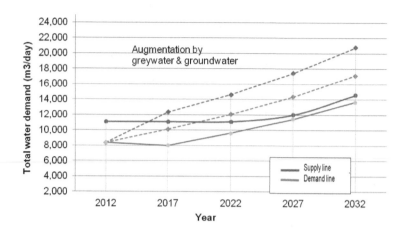

Figure 5.35 Demand and supply balance for minimum rate of water demand increase and augmentation by greywater

5.4.5 Comparison of the results to the World Bank report

The case study results obtained from this research are compared with the results by the World Bank (2012). Comparative results are presented in Table 5.17. As shown, the results with single values are from the World Bank report (2012) and values with mean and 95% interval are from this research. For example, for Cluster 1, the population forecasted in the World Bank report is 23,020, whereas from this study is [18,939, 22,862, 27,426] which means average population will be 22,862 with 95% confidence interval of 18,939 and 27,426.

The comparison of the results shows that the World Bank study results are within the range of the 95% confidence interval of this analysis. However, greywater and blackwater quantity of this analysis is relatively more than that estimated by the World Bank study. This is because the World Bank analysis considered the efficiency of these systems at a fixed value of 85 percent, whereas our analysis is based on the distribution functions and uncertainty sampling.

In contrast to the World Bank (Jacobsen et al., 2012), this research fully acknowledges the uncertainty associated to the data used for Mbale case study. It treats the associated uncertainties with proper description of the data, propagation through the simulation models, and systematic presentation of the results. The results presented with the uncertainty envelope and descriptive probabilistic information such as mean, standard deviation, and confidence levels can be used for further risk analysis. Moreover, the scenarios developed to analyse the future water demand, supply analysis, and select the best strategies considering the most probable confidence interval will be more realistic compared to description of the future.

Table 5.17 Comparative of analysis results of this study with the World Bank report

Cluster	Population	LPCD	Water leakage (m3/day)	Greywater (m3/day)	Blackwater (m3/day)	Total water demand (m3/day)
Cluster 1	23020	68.50	15%	691	442	1813
	[18939, 22862, 27426]	[52.5, 64.48, 87.61]	[173, 265, 369]	[499, 782, 1078]	[405, 686, 999]	[1481, 1811, 2148]
Cluster 2	14008	67.50	15%	402	265	1087
	[11524, 13911, 16688]	[54.2, 69.07, 85.44]	[107, 163, 228]	[310, 497, 695]	[237, 408, 604]	[900, 1118, 1348]
Cluster 3	11918	62	15%	295	195	850
	[9805, 11836, 14120]	[48.8, 62.9,79.6]	[83, 126, 173]	[222, 349, 486]	[223, 341, 481]	[728, 861, 1005]
Cluster 4	16144	68.50	15%	484	310	1272
	[13281, 16032, 19233]	[52.7, 64.9, 90.2]	[123, 188, 263]	[358, 574, 802]	[406, 471, 697]	[1038, 1288, 1553]
Cluster 5	27155	64.50	15%	735	477	2014
	[22340, 26968, 32351]	[49.3, 65.4, 84.5]	[195, 299, 418]	[537, 866, 1218]	[472, 789, 1142]	[1642 2042, 2459]
Cluster 6	126037	68	15%	3168	1720	9856
	[103690, 125170, 150160]	[51.4, 68.90, 90]	[933,1465, 2077]	[2829, 4558,6407]	[1886, 3543,5373]	[7877, 9992, 12263]
Total	218282	66.50	15%	5775	3409	16892
	[179582, 216778, 260052]	[51.5, 65.90, 86.9]	[1614,2506, 3528]	[4755, 7626, 10686]	[3629,6238, 8154]	[12938,17116, 20777]

Table 5.18 Comparative of natural water resources analysis results with the World Bank report

Natural sources	The World Bank report	This analysis
i) Groundwater	38.00 m3/day	Fuzzy membership function of [25, 30, 35, 40] m3/day After conversion to probability, mean of 37.12 m3/day and STD of 5.28 m3/day 95% confidence interval [28.64, 45.20] m3/day
ii) Nabijo and Nabiyonga Rivers	2000-5000 m3/day	Mean of 7565.34 m3/day and STD of 1840.90 m3/day 95% confidence interval [1719, 5341] m3/day
iii) Namatala River	2000-5000 m3/day	Mean of 3538.34 m3/day and STD of 1104.50 me/day 95% confidence interval [1724 , 5345] m3/day
iv) Manafwa River	5000-10000 m3/day	Mean of 7565.34 m3/day and STD of 1840.50 95% confidence interval [4528, 10570] m3/day
Total Supply	(90038-20038) m3/day	Mean 14680 m3/day and STD of 2384 m3/day 95% confidence interval [10665, 18479] m3/day

5.5 Discussion and Conclusions

This application example has demonstrated the applicability of developed methodologies particularly risk assessment framework, hybrid approach to uncertainty analysis, and multi-criteria analysis for the long-term decision making in UWS planning in Mbale, Uganda.

The analysis result showed that annual rate of population growth in Mbale will be 3.48% with STD of 0.57. This growth results total population of 216,778 with 95% confidence interval of [179,582, 260,052]. The future water demand for Mbale in year 2032 was found to be 17,116 m3/day with 95% confidence interval of [12,938, 20,777] m3/day. The total daily natural water resources for the town are 14,680 m3/day. Based on this analysis, there will be a water deficit of about 2,435.19 m3/day, about 17 percent. Assuming average rate of water consumption 65.90 LPCD, it will affect about 36,953 people in 2032. The risk level from this classification lies within the lowest level, however, water supply must cover 100 percent of the population, challenging for NWSS and Mbale.

In order to respond to the identified risk, available water resources and local resources including groundwater and recycling water were assessed. The information about river sources and groundwater is very limited. Both probabilistic and fuzzy set theory were applied to describe the uncertainty based on the available information. For example, the safe yields from the rivers were considered probabilistic whereas the groundwater was represented by fuzzy membership function. The propagation of uncertainty was performed by LHS after transformation of fuzzy into PDF by the transformation technique developed from this research. The output was presented with distribution and 95% confidence interval.

Three alternative potential sources of water were identified for the analysis. Selection of water resources is a multi-criteria analysis. The multi-criteria selected in environmental, technical, economic, and social dimensions were applied using the fuzzy based MCDM framework. The evaluation index result showed that the recycling of water (greywater) has the least priority for future development. This could be due to social and economic factors as its cost of production is very high and the social acceptability for this option was very poor.

In order to address the calculated risk, three scenarios were developed to meet the water demand from the different sources. The scenarios were developed within the 95% confidence interval of the analysis results. It is inevitable that Mbale needs to go for the reuse and recycle option unless the groundwater sources can be found. Although the scenarios show the achievable strategy, it needs further detailed analysis of costs, which is beyond the scope of this analysis.

The result shows that this analysis is more comprehensive, scientific, and provides more information than the World Bank report. This analysis fully recognises and acknowledge the uncertainty associated to the for the future and data scarce situations. After careful comparison, it can be seen that the recommendations made by the World Bank report for future water resources assessment are very conservative. The report recommends building dams in different rivers. However, it analyses the supply demand balance only considering dry flow. The dam cannot be chosen and recommend without assessing the annual water resources flow. This case study reveals the importance of uncertainty analysis and implication in decision-making in the cases of data scarce situation like Mbale.

This case demonstrates the developed framework, hybrid approach to uncertainty analysis, and risk assessment for a decision-making process. However, numbers of input parameters used in these cases are not extensive. The analysis lacks analysis of different types of distributions functions and alternative shapes of the membership functions as the input parameters and their effect on the results. The analysis could have been more interesting if it was tested in any complex model with multiple inputs.

Chapter 6   Conclusions and Recommendations

This Chapter presents conclusions drawn from this research and suggests recommendations for strengthening this research in the future.

6.1   Conclusions

The main goal of this research was to develop a framework, methodology, and set of tools for decision-making under risk and uncertainty for sustainable development of urban water systems. The specific objectives were:

- To develop a risk assessment framework for urban water systems considering future change pressure and their associated uncertainties.
- To develop a hybrid modelling method that allows uncertainties to be described and propagated through urban water models.
- To develop a multi criteria decision-making framework that is tailored for UWS operating under uncertainty.
- To demonstrate the developed frameworks and methods by applying them in different real cases of UWS.

The research objectives were addressed through a theoretical and modelling approach and applications in case studies in three cities.

6.1.1   A new risk assessment framework for urban water systems considering uncertainties

This research has proposed a new risk assessment framework for analysing risk, uncertainty, and decision-making in UWS experiencing global change pressures. The framework combines a hybrid approach to uncertainty analysis, fuzzy set theory based risk assessment technique, and a fuzzy set theory based hierarchical multi-criteria analysis method.

The hybrid approach to uncertainty analysis developed from this research allows the handling of both aleatory and epistemic uncertainty in UWS within a single framework. This approach is different from other as it firstly transforms uncertainty into a single type and propagates it homogenously. Algorithms for the transformation are based on evidence theory.

In the framework, the risk in UWS is analysed using a fuzzy set theory based semi-quantitative approach. The semi-quantitative approach allows the handling of likelihood and severity consequences in both a fuzzy form and quantitative terms, which is new in UWS modelling. This method helps to demonstrate the objective as well as subjective risk results.

Finally, a hierarchical framework of a MCDM is included in the risk assessment framework. The MCDM framework is based on the fuzzy set theory with AHP weighting schemes. It allows a comprehensive analysis of system performance considering different types of uncertainties. This is applicable for performance based planning as well as risk management by selecting the best strategy or option to respond the risk in a complex system.

The framework provides a comprehensive view of risk and uncertainty analysis for a sustainable urban water systems analysis and it can be applicable for other sectors without much modification. The application of the risk assessment framework in case studies demonstrates it's applicability for real-world urban water systems.

6.1.2 Hybrid method of uncertainty modelling to describe and propagate uncertainty in UWS

This research has developed a new hybrid approach to uncertainty analysis. This consists of (i) describing the uncertainty by the PDFs and MFs, (ii) transforming the uncertain information into the probabilistic or fuzzy forms by the application of the evidence theory based transformation algorithms, (iii) homogenously propagating the transformed uncertainty through any probabilistic or possibilistic technique of uncertainty analysis and (iv) statistical analysing of uncertainty results. Thus, the description, transformation, propagation, and result analysis are the main steps for the hybrid approach of uncertainty analysis. The developed hybrid approach can be applied to any complex modelling in UWS and other systems, which has multiple types of uncertainties in their inputs or parameters used for the modelling.

Description of the uncertainty and identification of the sources of uncertainty is important steps to select the methods of uncertainty analysis. Three sources of uncertainty classified from this study are natural, technical, and social. Natural sources are uncertainties related to climate change and changes in climatic parameters. Technical sources of uncertainties are due to system models, system interactions, and operations. Social sources of uncertainties stem from human-induced decision-making process, which are subjective, qualitative, or vague. Each source could have multiple types of uncertainty, and that exist in input parameters and system model. Aleatory uncertainty stems from the variability within known populations and represents the inherent randomness of events. This type is represented by a probability distribution function or a probability mass function. Epistemic uncertainty results from a lack of information and is, therefore, prone to subjectivity, ambiguity, and ignorance. Epistemic uncertainty is represented by fuzzy membership functions.

This research developed algorithms for transforming probabilistic information into fuzzy forms and vice versa, based on the Dempster-Shafer theory. The transformation ensures that the major principles of transformation such as consistency and preference are preserved. The technique is simpler than other existing methods and applicable to all types of distribution and membership functions used in UWS modelling.

The decision for transforming probabilistic information into fuzzy forms and vice versa depends on the numbers of input parameters, types of uncertainty, and objectives of the analysis. If the majority of the data is qualitative and imprecise or data are scarce, then the analysis results will correspondingly be less precise. The possibility is an ordinal and epistemic scale of uncertainty. Uncertainty values represented by possibility distribution functions are less precise than those of the probability. Thus, conversion from possibility to probability could lead to derive less precise probabilities from less precise data sources. In such cases, a fuzzy set approach will be applied. In contrast, the probabilistic approach will be applied if the majority of the input data are probabilistic. It is noted that probability is always sensitive to noise and errors. Small errors of prior probabilities may produce wrong reasoning results. It requires sensitivity analysis considering the sources of uncertainty, types of uncertainty and number of input parameters.

The homogenous propagation of uncertainty is achieved by using probability or fuzzy set based technique. The selection of any technique depends on the information available, complexities of a model and objectives of the analysis. This research applied MCS, LHS, and Bootstrapping techniques and fuzzy alpha-cut technique for propagation. These techniques are widely used and robust enough for uncertainty propagation. Additionally, algorithms for modelling the selected techniques have been presented with examples. For the transformed probabilistic uncertainty, random samples are taken from the CDF generated out of PDFs. The type of PDFs can also be selected by evaluation of goodness-of-fit of a probability distribution model, such as the chi-square test, Kolmogorov-Smirnov Test or Anderson-Darling Test for parametric distributions.

The results of the uncertainty analysis are presented in distribution forms with descriptive statistical information such as mean, standard deviation and 95% confidence interval. This is useful for further risks assessment and decision-making for the risk management.

The hybrid methodology was applied for water demand analysis in Birmingham, UK. The result of the analysis was compared with other hybrid and homogenous uncertainty analysis approaches. The results indicate that this approach performed better compared to the other hybrid approaches for the following reasons: (i) it allows for uncertainty propagation using either a probability approach, such as MCS or LHS, or possibilistic approach using the fuzzy alpha-cut- technique which are well known and well tested techniques of uncertainty propagation, (ii) the results are more reliable (i.e., having less uncertainty range) compared with the other heterogeneous approaches, and (iii) the method is very easy to analyse (do not require additional mathematical complexities to process and post process the results) and applicable in any level of complexity.

6.1.3 Multi-criteria decision making framework tailored for UWS

A fuzzy set theory based MCDM method for UWS operating under uncertainties is developed from this research. The method is very simple, capable of aggregating the different criteria, flexible to capture the uncertainty, and allows stakeholders and decision-makers to incorporate their preferences and priorities.

The proposed MCDM framework is hierarchical in order to analyse the performances of various criteria and sub-criteria at different levels under different dimensions (i.e., technical, social, environmental, economic, etc.). The indicators measure the quality or performance of a component under any criteria or dimension. The indicator values are classified in different grades based on best practices, experts' consultation and in consideration of local conditions.

The relative importance of each criteria and indicator is decided by applying AHP technique. The AHP technique is widely used, transparent and user friendly. Stakeholders and experts can be engaged in the decision making process by determining the types of indicators to be used and ranking each of the indicators and their criteria.

Fuzzy Synthetic Evaluation Techniques used for the aggregation of the criteria can capture different types of uncertainties. It can represent crisp, probabilistic, and qualitative data. The performance or preferences are aggregated from lower level to upper level by fuzzy operations. The results are presented both in fuzzy form and in an index form after fuzzification and defuzzification processes. Therefore, the aggregation operation consists of numerous numbers of fuzzification, matrices operation and defuzzification. The results are easy to understand and compare the different components of a system or strategy with each other. This technique is thus capable of capturing data, which are ambiguous and linguistic, addressing the uncertainty and presenting the results in a simple form.

The developed framework and method was demonstrated in a performance analysis of urban infrastructure system in Kathmandu, Nepal. The method demonstrated on how an integrated approach can be operationalise for urban infrastructure planning systems to optimize the investment costs and the maximize the overall performance in the systems.

6.1.4 Application of the developed frameworks and methods in UWS

Analysis of risk and uncertainty of future water availability in Birmingham

The application case of Birmingham, UK for water availability analysis under uncertainty in 2035 presented in Chapter 3 demonstrated the applicability and benefit of the hybrid approach of uncertainty analysis and risk assessment in strategic planning and decision-making.

The case study modelled and analysed the future water demand in Birmingham in the year 2035 considering the uncertainty in the micro-demand of water. The uncertainties considered for the analysis are from the future population growth, climate change, socio-economic change, and water loss by leakage. Both the probabilistic theory and fuzzy set theory were used to describe the uncertainty. The uncertainties were represented by PDFs and MFs.

The water demand was analysed considering both climatic and non-climatic factors. The climatic demand is uncertain; however, its impact in water demand will be negligible compared with the non-climatic demand. The main drivers for the future water demand of Birmingham will be the domestic demand. The industrial demand is expected to be similar (as many industries were moved out from Birmingham since 1990s); however, it is equally uncertain like leakage. The socio-economic analysis shows less likely changes in social structure in the future and continuation of the existing water demand in different ACORN groups. It will have a minimal impact to the overall future demand.

The current water resource capacity of the STWL Birmingham zone is 314.90 ML/day (which is greater than the STWL prediction of 296.50 ML/day). The analysis showed that the water shortage rates in Birmingham will be @ 2.58% (i.e., 8.15 ML/day), and 14.24% (i.e., 44.85 ML/day), for the mean, and maximum growth projections of the uncertainty envelope, respectively. This level of water scarcity will affect 71,275 people assuming 114.36 LPCD by 2035.

The hybrid approach to uncertainty analysis developed from this research was applied for uncertainty analysis. The uncertainties were propagated using the algorithms for LHS, MCS, and moving-block bootstrapping and fuzzy alpha-cut techniques. The result was compared with the uncertainty technique proposed by Guyonnet et al. (2003). The range of uncertainty obtained from the proposed method was found to be more precise than the result from Guyonnet et al. (2003) algorithm.

The comparison of the results showed that the proposed hybrid method is easy for modelling a large system. It does not require any separate mathematical operations for the description, propagation and output analysis.

Performance based urban infrastructure planning in Kathmandu Nepal

The developed MCDM framework and method was applied for the overall performance assessment of urban infrastructure system in Kathmandu City (KMC). The application example demonstrated in Chapter 4 showed the potential benefit of the developed framework for a complex system and the benefit of capturing multiple sources of information and uncertainty. It also showed how the decision makers and stakeholders can be engaged while selecting the performance indicators, classifying performance levels (i.e., poor to excellent level), and assigning the relative importance of the indicators. The fuzzy set theory based MCDM methods were found to be capable of capturing both quantitative and qualitative data. The AHP based priorities setting approach allowed stakeholders to engage in and reflect on the decision-making process.

The result showed that the overall performances of the urban infrastructure systems in KMC are in poor condition, and the performance index values of the independent systems were within a similar range. The reasons behind the poor performances are poor reliability and effectiveness (meaning the infrastructure systems are not adequate), inadequate level of services, and low social and environmental quality. The analysis results presented in Chapter 4 demonstrated that how decision-makers can plan to improve the overall quality of infrastructure services in Kathmandu by investing in critical infrastructure systems.

The application of the multi-criteria decision-making framework in Kathmandu illustrates that the framework can be applied in complex decision analyses of complex and interdependent systems either for performance based planning or risk management.

Risk of water scarcity analysis and risk management under uncertainty in Mbale, Uganda

The application example in Mbale, Uganda presented in Chapter 5 demonstrated the combined applicability of all developed methodologies particularly the risk assessment framework, the hybrid approach to uncertainty analysis and the multi-criteria analysis for strategic decision-making in water resources planning in a data limited case.

The future water demand and supply were forecasted by considering the uncertainties associated with the rate of annual population growth, river water discharge, safe groundwater yield, and the rate of water leakage from the water supply systems. The leakage and groundwater sources are represented by fuzzy membership functions. Other parameters were described using PDFS. During data analysis of the study, particularly for the fuzzification of the uncertain data and assignment of weights by the AHP technique, water experts working at NWSC, Uganda, engineer and sociologist working in the municipality of Mbale, Uganda, and researchers involved in the World Bank study project from the University of South Florida were consulted. The hybrid approach to uncertainty analysis processed the uncertainty by transforming the FMFs into PDFs. The reason for transforming fuzzy form to the probabilistic form is because the majority of the data are best described by PDFs such as population growth, water consumption rate and river water flows.

The analysis results showed that the average future water demand for Mbale in year 2032 will be 17116 m3/day. The total average daily natural water resources available for the town are 14680 m3/day. Based on this analysis, there will be a water deficit of about 2435.19 m3/day, which is about 14.22 %. Assuming a 70 LPCD water consumption rate, the water scarcity will affect about 34,788 people in 2032. The risk level from the classification lies within the very low level. Covering 100% population by water supply service will be challenging for NWSC.

The fuzzy set theory based MCDM framework was applied to select the best source from three potential sources. Environmental, economic, technical, and social criteria were used for measuring the preference levels. The relative importance of each of the indicators and criteria were assigned using the AHP technique. The analysis results showed the relative importance of three sources for Mbale. The preference of greywater reuse was found to be least due to social and economic reasons.

In order to address the calculated risk, three scenarios were developed to meet the water demand from the various sources. Based on the analysis it is inevitable that Mbale goes for reuse and recycle of wastewater option unless additional groundwater sources could be found. The scenarios show that the optimal achievable strategy for Mbale must use recycled water in order to meet the future water demand by 2027.

The results of the analysis based on the proposed risk assessment framework were compared with the results of the conventional method as applied in the World Bank (2012) report. The results of the mean values of total water demands were comparable; however, there are differences in the recommendations for the selection of water sources. This approach simulates the potential thousands of samples and generates uncertainty envelope. The uncertainty envelope with a certain degree of confidence can be used for further analysis such as magnitude of regrets and rewards from different options. The application of the proposed risk assessment framework in Mbale Uganda showed the utility of the framework uncertainty analysis in decision-making for the future and for analyses that are based on the several assumptions and limited data.

6.2   Recommendations

This research has developed frameworks, methodologies, and tools for decision-making under risk and uncertainty. While the thesis has addressed all the research objectives presented in Chapter 1, clearly many unexplored areas could be researched further to strengthen this field. Some of the recommendations for future research include.

- As discussed in the thesis, there are many major change pressures affecting urban water systems and these change pressures have a great deal of uncertainty associated with their prediction and impact analysis. This thesis has focussed on how to capture and articulate uncertainties by understanding the types and sources of uncertainties. However, greater research is required on how to minimize these uncertainties by improving their predictions.

- The thesis presented a MCDM framework for urban water systems operating under uncertainty. This MCDM framework could be further strengthened by incorporating it into an optimisation framework. By doing so, this would allow decision makers to identify optimal design and control strategies that both minimize costs while maximizing the benefits of the water system. It is envisaged that evolutionary algorithms could be utilised along with the proposed hybrid method and urban water simulation models.

- From the outset, the thesis argued that urban water systems are experiencing several future change pressures and that these pressures have a high degree of uncertainty. In fact, the urban related forecasts are extremely poor in respect to the expected operational life span of many urban water systems (40 to 100 years). Hence, the design of urban water systems should recognize the inevitable uncertainties and move to the design of more flexible and adaptive systems. These adaptable systems should have the capacity to act and respond on alterations of system performances in efficiently, timely, and the cost-effective way. Research is required on how to apply the developed frameworks so that they can contribute to the new emerging area of flexible design.

- It is important to understand the risk assessment of urban water systems within the context of integrated urban resource management, where all resource flows in a city, such as water, energy, and transport are considered. To achieve this, integrated models need to be developed that understand and articulate the flows and relationships between the different infrastructure systems. The integrated models will allow articulates the impact of a failure of one infrastructure system to another (e.g. impact of power failure on pumping in a water system). This thesis has attempted to describe this concept through the case study on integrated urban infrastructure planning in Kathmandu, Nepal (Chapter 4). However, greater research is required on the development of an integrated risk assessment framework, tailored for complex urban systems.

- It has been well recognised that many future change pressures will impact UWS. However, due to interdeterminacy of future change pressures, in many cases it is not possible to predict and estimate uncertainty accurately. Thus, in order to develop sustainable and resilience urban water infrastructure system, there is a need to develop design principles based on the adaptive and resilience concept.

- There are wide applications of Bayesian approach to uncertainty analysis. There is equal possibility of including the Bayesian approach in the hybrid approach; however, this will require further work in the future.

- Describing uncertainty is first steps in uncertainty analysis. Selection of particular PDFs can be archived by using best-fit and statistical analysis. However, research on selection of a shape of the membership functions is still in early stage. This will require future research to select the best shape to represent the particular types of information.

APPENDIX I

PROBABILITY AND PROBABILITY DISTRIBUTION FUNCTION

Classical definition of probability

The classical probability theory relates to the situation where an experiment can be repeated indefinitely under identical conditions, but the observed outcome is random. Empirical evidence suggests that the relative occurrence of any particular event, which is a relative frequency, converges to a limit according to the increase in the number of repetitions of the experiment. This limit is called the probability of the event. The probability of an event is the ratio of the number of favourable outcomes (simple events) of an experiment to the total number of outcomes in the sample space. If the probability of an event A is $P(A)$ and X is the universal set, the three axioms of probability in this information are:

Axiom 1: The probability of an event is always non-negative.

$$P(A) \geq 0 \qquad (I.1)$$

Axiom 2: The probability of the universal set is 1.

$$P(X) = 1 \qquad (I.2)$$

If a variable X, which has a set of possible discrete states $x_1 . \ldots .. x_n$. Then the probability of variable X being in state x_i is denoted by $P(X = x_i)$, and it follows from the axioms that:

$$\sum_{i=1}^{n} P(A_i) = 1 \qquad (I.3)$$

Axiom 3: If two events E_1 and E_2 are mutually exclusive, the probability of their union is equal to the summation of their probability and the probability of the universal set is 1.

$$P(A_1 \cup A_2) = P(A_1) + P(A_2) \qquad (I.4)$$

In general, $P(A_1 \cup A_2) = P(A_1) + P(A_2) - P(A_1 A_2)$ \qquad (I.5)

The Bayesian approach to probability

Probability theory is also applied to situations in which as frequency is not its source, or repeated event, however event is expressed as a degree of belief, called the *subjective* or *Bayesian interpretation*. In the subjective view, an event is a statement, and the (subjective) probability of the event is a measure of the degree of belief. The basic idea in the application of this approach is to assign a probability to any event on the basis of the current state of knowledge, and to update it in the light of the new information.

Conditional probability

Conditional probabilities are essential to a fundamental rule of probability calculus. The probability of joint occurrence of n independent events $(A, B, \ldots \ldots N)$ is the product of their individual probabilities. That is,

$$P(A \cap B \cap C, \ldots \ldots \cap N) = P(A) * P(B) * \ldots P(N) \tag{I.6}$$

The probability of joint occurrence of n dependent events $(A, B, \ldots \ldots N)$ is obtained from,

$$P(A \cap B \cap C, \ldots \cap N) = P(A) * P(B/A) * P(C/A \cap B) .. P(N/P(A \cap B \cap C \ldots \cap N_{n-1}) \tag{I.7}$$

where $P(C/A \cap B)$ denotes the conditional probability of $P(C)$ given the occurrence of other $P(A)$, $P(B)$ are possible.

Bayesian statistics are associated with the calculation of conditional probabilities. The basic version of Bayes' rules is,

$$P(hypothesis/evidence) = [P(evidence/hypothesis) P(hypothesis)]/P(evidence) \tag{I.8}$$

where, $P(evidence/hypothesis)$ - is a likelihood *function*, and $P(hypothesis)$ is the *prior*.

Alternatively, Bayes theorem conditioned on C can be written as:

$$P(B/A,C) = \frac{P(A/B,C)P(B/C)}{P(A/C)} \tag{I.9}$$

The left hand-side term, $P(B/A,C)$, is the *posterior* probability and it gives the probability of the hypothesis B after considering the effect of evidence A in the context C. The $P(B/C)$ term is just the *prior* probability of A given C alone; that is, the belief in A before the evidence B is considered. The term $P(A/B,C)$ is called the likelihood, which gives the probability of the evidence assuming the hypothesis B and background information C is true. The denominator of the right-hand term $P(A/C)$ is the prior probability of the evidence that can be regarded as a normalising or scaling constant.

Probability distribution

In probability theory, discrete random variable is represented by a *probability mass function* (Figure I.1), whereas a continuous random variable is represented by *probability density function* (PDF) $p_X(x)$ (Figure I.2). *Cumulative distribution function* (CDF) or simply *distribution function* (DF) $P_X(x)$ is the cumulation of the probability distributions. Any function representing the probability distribution of a random variable must necessarily satisfy the axioms of probability. Some of the commonly used PDF in uncertainty analysis are: uniform, triangle, normal, lognormal, and exponential models based on the information it contains.

The probability density function of continuous random variables is given by,

$$P_X(b) - P_X(a) = P(a < X < b) = \int_{-\infty}^{b} p_X(x)dx - \int_{-\infty}^{a} p_X(x)dx = \int_{a}^{b} p_X(x)dx \qquad (\text{I.10})$$

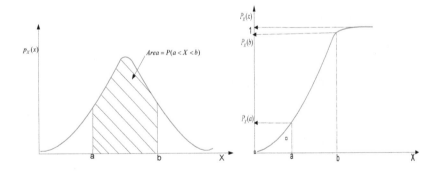

Figure I.1 (a) Probability density function (PDF) and (b) cumulative distribution function (CDF)

Similarly, the probability density function of continuous random variables is given by,

$$P(a < X \leq b) = \sum_{all\ x_{i \leq b}} p_x(x_i) - \sum_{all\ xi_{\leq a}} p_x(x_i) \qquad (\text{I.11})$$

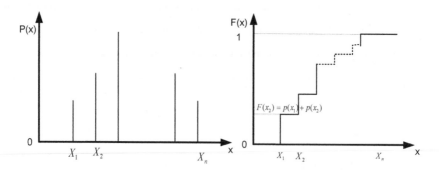

Figure I.2 (a) Probability mass function (PMF) and (b) cumulative distribution function (CDF)

APPENDIX II

FUZZY SETS AND FUZZY ARITHMETIC

Definitions on fuzzy sets

The concept on fuzzy set theory has been introduced in Chapter 2. More detailed coverage of this topic can be found in Bardossy and Duckstein (1995), Dubois and Prade (1988), Kaufmann and Gupta (1991), Ross (2008) and Zimmermann (1991).

Fuzzy set

If X be a universe set of x variables, then Ã is called a fuzzy (sub) set of X, if $\tilde{A}$ is a set of ordered pairs:

$$\tilde{A} = \left\{ (x, \mu_{\tilde{A}}(x)); \ x \in X, \ \mu_{\tilde{A}}(x) \in [0,1] \right\} \tag{II.1}$$

where $\mu_{\tilde{A}}(x)$ is the grade of membership of x in $\tilde{A}$.

The function $\mu_{\tilde{A}}(x)$ is the *membership function or grade of membership* (also degree of compatibility or degree of truth) of x in $\tilde{A}$. A membership function (Figure II.1) maps every element of the universal set X to the interval [0, 1], i.e.,

$$\mu_{\tilde{A}}(x): X \to [0,1] \tag{II.2}$$

202

Membership function

As shown in Figure II.1, a membership function (MF) is a curve that defines how each point in the input space is mapped to a membership value (or degree of membership) between 0 and 1. Membership is also defined as the *degree of belief*, also called *belief level*, to which the element belongs to the set. The membership may take any value between and including 0 (no membership) and 1 (full membership). The input space is sometimes referred to as the universe of discourse.

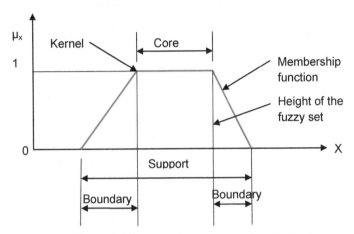

Figure II.1 Features of a normal membership function

Height of a fuzzy set

The *height* of a fuzzy set is the maximum value of the membership in its membership function, i.e. the height of the fuzzy set $\tilde{A}$

$$h(\tilde{A}) = \max\{\mu_{\tilde{A}}(x_1),......,\mu_{\tilde{A}}(x_n), x \in X\}$$ (II.3)

Support of a fuzzy set

The *support* of the membership function of a fuzzy set is the region of the universe that is characterised by nonzero membership in the fuzzy set, i.e.

$$Sup(\tilde{A}) = \{x \mid \mu_{\tilde{A}}(x_n) > 0\}$$ (II.4)

Normal and subnormal fuzzy set

A fuzzy set which has at least one element with unity membership is called a *normal* fuzzy set, that is, there exists $\mu_{\tilde{A}}(x) = 1$ (for some $x \in X$. If the height of a fuzzy set is less than unity the fuzzy set is called a *subnormal.*

Convex and non-convex fuzzy set

A *convex* fuzzy set the membership function is monotonically increasing or decreasing. The membership function in a convex fuzzy set is either (i) monotonically increasing, or (ii) monotonically decreasing, or (iii) monotonically increasing and monotonically decreasing with increasing values of the elements in the universe. For every real number, say x_1, x_2 and x_3, with $x_1 < x_2 < x_3$.

Such that, $\mu_{\tilde{A}}(x_2) \geq \min\{\mu_{\tilde{A}}(x_1), \mu_{\tilde{A}}(x_3)\}$

(II.5)

If the membership function is not monotonically increasing, decreasing, or both monotonically, increasing or decreasing with increasing value of element in the universe it is called *non-convex fuzzy set.*

Fuzzy number

A fuzzy set which is normal and convex is called a *fuzzy number.*

Operation of fuzzy sets

Similar to classical sets operation, *union, intersection,* and *complement* are the three basic operations in fuzzy sets. Given two fuzzy sets $\tilde{A}$ and $\tilde{B}$ the union, intersection, and complement operations are represented as (Figure II.2).

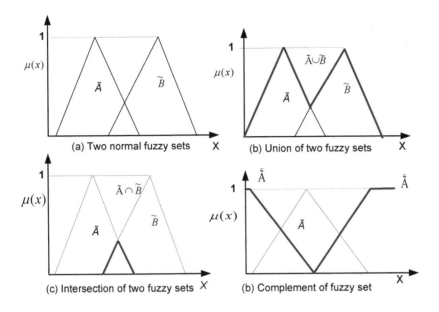

Figure II.2 (a) Fuzzy sets $\tilde{A}$ and $\tilde{B}$ represented by triangular membership functions with main operations: (b) union, (c) intersection, and (d) complement.

Union: $\mu_{\tilde{A}\cup\tilde{B}}(x) = \mu_{\tilde{A}}(x) \vee \mu_{\tilde{B}}(x) \Rightarrow \max(\mu_{\tilde{A}}(x), \mu_{\tilde{B}}(x)), \; x \in X$ (II.6)

Intersection: $\mu_{\tilde{A}\cap\tilde{B}}(x) = \mu_{\tilde{A}}(x) \wedge \mu_{\tilde{B}}(x) \Rightarrow \min(\mu_{\tilde{A}}(x), \mu_{\tilde{B}}(x)), \; x \in X$ (II.7)

Complement: $\mu_{\tilde{A}}(x) = 1 - \mu_{\tilde{A}}(x)$ (II.8)

Fuzzy Arithmetic

Fuzzy arithmetic is based on the fuzzy extension principles introduced by Zadah (1975) and elaborated by Yager (1986). Fuzzy number is generally considered as generalisation of interval of confidence. The operation varies according to the monotonic and non-monotonic functions.

For a non-monotonic function, its implementation requires an algorithm for the determination of maximum and minimum values of the function to be evaluated.

Interval of arithmetic operation

If $[a,b]$ and $[c,d]$ are any two intervals, the arithmetic operation is performed in the following way

Addition:

$$[a,b]+[c,d]=[a+c,b+d]$$

(II.9)

Subtraction:

$$[a,b]-[c,d]=[a-d,b-c]$$

(II.10)

Multiplication:

$$[a,b]\times[c,d]=[\min(ac,ad,bc,db),\max(ac,ad,bc,db)]$$

(II.11)

Division

$$[a,b]\div[c,d]=[\min(a/c,a/d,b/c,d/b),\max(a/c,a/d,b/c,d/b)]$$

(II.12)

Power:

$$[a,b]^{[c,d]}=[\min(a^c,a^d,b^c,d^b),\max(a^c,a^d,b^c,d^b)]$$

(II.13)

Operation of Triangular fuzzy number

The membership function μ of a triangular fuzzy number $\tilde{A}$ on a range $(-\infty,+\infty)$ is given by

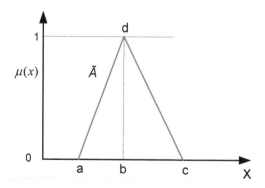

Figure II.3 Triangular fuzzy number

$$\mu(x) = \begin{cases} \dfrac{x-a}{b-a} & a < x < b \\[2mm] \dfrac{c-x}{b-a} & b < x < c \\[2mm] 0 & otherwise \end{cases}$$

(II.14)

where $a < b < c$ and a and c lower and upper values of the support of the fuzzy and b is the most likely value.

For any two triangular fuzzy numbers, $\tilde{A}_1$ and $\tilde{A}_2$ with the known three values, the arithmetic operations are performed in the following way

Addition: $\tilde{A}_1 + \tilde{A}_2 = (a_1 + a_2, b_1 + b_2, c_1 + c_2]$ (II.15)

Subtraction: $\tilde{A}_1 - \tilde{A}_2 = (a_1 - c_2, b_1 - b_2, c_1 - a_2]$ (II.16)

Multiplication: $\tilde{A}_1 \times \tilde{A}_2 = (a_1 \times a_2, b_1 \times b_2, c_1 \times c_2]$ (II.17)

Division: $\tilde{A}_1 \div \tilde{A}_2 = (a_1 / c_2, b_1 / b_2, c_1 / a_2]$ (II.18)

Power: $\tilde{A}_1^{\tilde{A}_2} = (a_1^{a_2}, b_1^{b_2}, c_1^{c_2}]$ (II.19)

Defuzzification methods

Defuzzification is the process of converting a fuzzy a fuzzy quantity to a precise quantity. There are various methods for defuzzificaiton, such as max-membership principle, centroid method, weighted average method, mean–max membership, centre of sums, centre of largest area, and first of maxima or last of maxima. Two methods commonly used in the defuzzification process are *centre-of-area method* or *centroid method* and *max-membership method or height method* (see, Ross, 2008).

Figure II.5 graphically represent both the method of defuzzification. The centre of area method is obtained by calculating the centre of gravity of the fuzzy membership function and is given by

$$x* = \frac{\int \mu_x(x) x \, dx}{\int \mu_x(x) \, dx}$$

(II.20)

In height method, the value corresponding to the maximum membership is considered as the defuzzified quantity and it is represented as,

$$\mu_x(x*) \geq \mu_x(x); \ x \in X$$

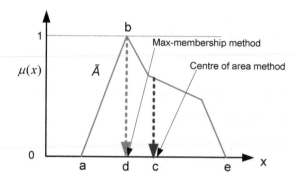

Figure II.4 Defuzzification by max-membership method and centre of area method

APPENDIX III

Risk Assessment and Modelling

Risk in a system is the combination of the probability of a performance failure, and its severity or consequences. The performance failure represents the failure of meeting the designed level of service, which is possible to calculate using a system model or some other analytical technique. The consequence accounts number of people affected, the economic losses, or the environmental impact. The consequence is difficult to predict and will be best dealt with by using the fuzzy set theory. In the case of a complex system, the total risk in of the complex system is the cumulative risk of each different sub-system under consideration. Thus, the risk at any level of a system (i.e., sub-system or system level) is aggregated by considering the probability of failure of performance and its associated severity, as in Equation 3.56.

$$R_j = \sum_{i=1}^{n}(P_{ij} \times S_{ij})$$
(III.1)

where R_j is the total risk in a system j; P_{ij} and S_{ij} are the probability and severity of failures, $i = 1,.....,n$. It is noted that Equation (3.56) requires a fuzzy operation.

Representing the likelihood (i.e., probability of failure), consequence and risk in an ordinal scale is very common in risk analysis. However, there are neither definitive assumptions nor analysis techniques for selecting the number of risk levels. While classifying the ordinal scale, experiments by psychologists, such as those of Miller (1956), suggest that the maximum number of chunks of information is on the order of seven, plus or minus two. With respect to this, it is often recommended that the number of categories be restricted to no more than seven (Karwowski and Mital, 1986). Normally, too few levels will be inadequate to represent the real knowledge of the analysts, while too many levels will add extra difficulties in assessment.

In practice, the WHO (2010), in their quality risk analysis, propose five levels of likelihood: (i) *Almost certain*: once a day; (ii) *Likely*: once per week; (iii) *Moderate:* once per month; (iv) *Unlikely:* once per year; (v) *Rare:* once every 5 years. Similarly, IPCC (2012) uses seven levels to indicate the assessed likelihood, described as (i) *Virtually certain:* 99-100% probability; (ii) *Very likely:* 90-100% probability; (iii) *Likely:* 66-100% probability; (iv) *About as likely:* as not 33-66% probability; (v) *Unlikely:* 0-33% probability; (vi) *Very unlikely:* 0-10% probability; (vii) *Exceptionally unlikely:* 0–1% probability.

In addition, the selection of a particular shape of the membership function depends on the type of information to be represented and its associated uncertainty. There are several methods for selecting an appropriate shape (see, Bilgic and Turksen, 1999). However, in many cases, the shape of the membership function does not influence the results of the engineering application (Klir and Yuan, 1995). Thus, this research adopts a triangular shape of fuzzy numbers to represent the risk parameters for its simplicity.

Probability of risk

Probability or likelihood of risk is a measure of the performance failures in UWS. The degree of likelihood will depend on how often and at what rate it affects the designed performance of the system. The designed performances are the expected minimum level of services for the systems, such as minimum depth of flow, quality and quantity of supplied water, pressures in the distribution systems, etc. The likelihood of failure is identified either from past observations or by using independent system models.

A system model can be used to calculate the performance of a system, such as pressure in a distribution system (Figure III.1). This calculated performance is then converted into the fuzzy membership function through the fuzzification process. Seven categories of linguistic representations are assumed to express the probability of performance failure and are represented by triangular shaped membership functions (Figure III.2, Table III.1). Any value from 0 to 1 will represent the relative magnitude of likelihood—for example, a value of 0 implies no impact on the system's performance, while a value of 1 implies complete failure of a system performance.

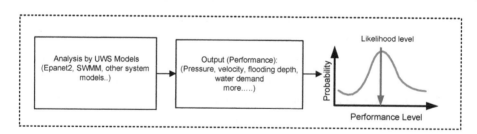

Figure III.1 Likelihood of risk

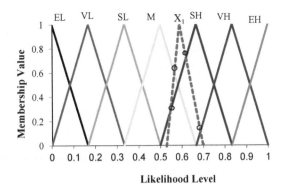

Figure III.2 Seven levels of likelihood for the risk calculation

Table III.2 Linguistic levels and explanation of likelihood of a risk

Linguistic representation	Likelihood level	Description
1) Extremely Low (EL)	[0, 1/6]	Impossible during the designed life of a system
2) Very low (VL)	[0, 1/3]	Very unlikely
3) Slightly low (SL)	[1/6, 1/2]	Unlikely
4) Medium (M)	[1/3, 2/3],	Likely
5) Slightly high (SH)	[1/2, 5/6]	Possible
6) Very high(VH)	[2/3, 1]	Quite possible
7) Extremely high (EH)	[5/6, 1]	Expected

Representing the likelihood of risk in a fuzzy form requires a fuzzy mapping operation. For example, if the likelihood of a performance failure has three values of measurement (i.e., $X_1 = 0.52$, 0.61 and 0.69), each corresponding to the minimum value, the most likely value and the maximum value respectively, and is represented by the dashed line triangle in Figure III.2, then X_1 is mapped on a fuzzy scale that intersects with numerous membership functions. If it has an intersection at more than one point, the *maximum operator* is used to determine the membership value (Klir and Yuan, 1995). As shown, X_1 intersects with the medium (M) membership function at 0.61 (i.e., $\mu_M = 0.61$); the slightly high (SH) membership function at two points, 0.78 and 0.28 (thus requires the maximum operator); and with the very high (VH) membership function at 0.18 (i.e., $\mu_{VH} = 0.18$). After the fuzzy *max operation* (i.e., with the maximum of 0.78 and 0.28) for the SH range, the SH fuzzy membership value of 0.85 (i.e., $\mu_{SH} = 0.78$) will result. Thus, the 7-tuple fuzzy set for the likelihood of risk under consideration will be [0, 0, 0, 0.61, 0.78, 0.18, 0]. Similar fuzzy mapping operations are applicable for crisp values.

211

Severity of failure

The *severity or consequence* of a failure measures the degree to which human life and environment will be impacted due to a system failure. Seven basic categories of severity levels are proposed in this thesis: *extremely low, very low, slightly low, medium, slightly high, very high,* and *extremely high* (Table III.2 and Figure III.3). These values are normalized with the dimensionless interval of [0, 1] such that the normalized severity level can be represented by fuzzy numbers defined in 0 and 1.

Thus, if S denotes the severity value and $S_{min}(=0)$ and S_{max} denote the possible minimum and maximum value of severities, respectively, then the normalized value will be

$$S \in [S_{min}, S_{max}] \Leftrightarrow S \in [0,1]$$
(III.2)

Table III.3 Linguistic levels and explanation of severity levels

Linguistic representation	Likelihood level	Description
Extremely low (EL)	[0, 1/6]	No impact to the users (human and environment)
Very low (VL)	[0, 1/3	Few users are affected
Slightly low (SL)	[1/6, 1/2]	A very small portion of users are affected
Medium (M)	[1/3, 2/3],	Part of the users are affected
Slightly high (SH)	[1/2, 5/6]	Many users are affected
Very high(VH)	[2/3, 1]	A large portion of the users are affected
Extremely high (EH)	[5/6, 1]	Most users are affected

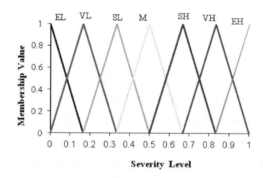

Figure III.3 Seven levels of severity for the risk calculation

212

Risk calculation

As discussed, the likelihood of risk and consequences are represented in fuzzy forms. The resulting risk in a system is calculated using the fuzzy alpha-cut operation (Figure III.4a). Five fuzzy numbers are considered to represent the resulting risk, which is characterized as very low, low, medium, high, and very high, as shown in Figure III.4(b) and Table III.3. The numbers closer to 1 have a proportionally higher level of risk.

Table III.4 Linguistic levels of risk level

Linguistic Variables	Risk Description	Likelihood Scale
1) Very low (VL)	Risk is acceptable	0% to 25%
2) Low (L)	Risk is tolerable but should be further reduced if it is cost effective	0% to 50%
3) Medium (M)	Risk must be reduced if it is reasonably practicable	25% to 75%
4) High (H)	Risk must be reduced	50% to 100%
5) Very high (VH)	Almost sure to occur	75% to 100%

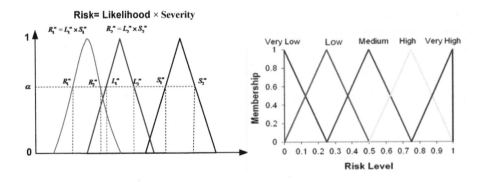

Figure III.4 (a) Calculation of risk at component level and (b) representation of risk

The magnitude of calculated risk will decide whether the induced risk in a system is acceptable or not. If the risk is unacceptable, different risk reduction options have to be analyzed to identify the best risk

213

reduction options. There are different guiding principles established in the literature to be considered during the risk evaluation (Christensen et al., 2003; Borja et al., 2006; McGill et al., 2007), for example:

- *Principle of as low as reasonably practicable (ALARP):* Risks that are clearly unacceptable must be reduced or eliminated under any circumstances. This includes the minimum level of services to be maintained in UWS, such as water quality, flood level, or water pressures level in a distribution system.

- *Principle of as low as reasonably achievable (ALARA):* Risks may be accepted if it is economically and/or technically unreasonable to reduce them. This will include, for example, a minimum flood water level that will appear rarely in a settlement.

- *Principle of reasonableness:* If risk is reasonable to reduce in economic and technical means, the risk shall be reduced regardless of the risk level. It covers the goals of local government, regulator and utilities who decide the level of services in their work area/city.

- *Principle of proportionality:* The overall risk resulting from an activity should not be unreasonably large compared to the benefits. This relates the cost and benefit to be considered while introducing a risk management measure.

- *Principle of allocation:* The allocation of risk in society should be reasonable/fair compared to how the benefits are allocated.

- *Principle of avoidance of disasters:* Risks with disastrous consequences should be avoided so that the consequences can be managed with readily accessible resources.

Calculation of Risk

In this thesis, the risk is calculated by fuzzy α-cut technique. However, this can also be calculated by fuzzy operation as suggested by Kaufman and Gupta (1991). In this operation, if likelihood of any hazard event and its corresponding severity is given by $\tilde{L} = (a_1, b_1, c_1)$ and $\tilde{S} = (a_2, b_2, c_2)$ respectively, then $\tilde{L}$ and $\tilde{S}$ can be represented as

$$\tilde{L}(x) = \begin{cases} \dfrac{x-a_1}{b_1-a_1} & a_1 < x < b_1 \\ \dfrac{c_1-x}{b_1-a_1} & b_1 < x < c_1 \\ 0 & otherwise \end{cases} \quad and \quad \tilde{S}(x) = \begin{cases} \dfrac{x-a_2}{b_2-a_2} & a_2 < x < b_2 \\ \dfrac{c_2-x}{b_2-a_2} & b_2 < x < c_2 \\ 0 & otherwise \end{cases} \quad (III.3)$$

214

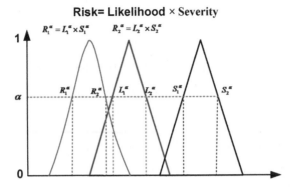

Figure III.5 Fuzzy operations for risk calculation

As we have defined the risk as the triple product of likelihood of risk, severity and vulnerability of a system or components. However, the vulnerability is represented by assigning the relative weight, which is scalar and crisp number obtained by AHP technique.

$$R_j = \sum_{i=1}^{n} (L_{ji} \times S_{ji})$$

(III.4)

The likelihood and consequences is represented by fuzzy membership functions and its fuzzy operation is undertaken according to alpha-cut operation (see Figure II.4).

$$\alpha = \mu_L(x) = \begin{cases} \dfrac{L^{\alpha}_1 - a_1}{b_1 - a_1} \\ \dfrac{c_1 - L^{\alpha}_2}{b_1 - a_1} \end{cases} \qquad \text{and} \qquad \alpha = \mu_S(x) = \begin{cases} \dfrac{S^{\alpha}_1 - a_2}{b_2 - a_2} \\ \dfrac{c_2 - S^{\alpha}_2}{b_2 - a_2} \end{cases}$$

(III.5)

Therefore,

$$\begin{cases} L^{\alpha}_1 = (b_1 - a_1)\alpha + a_1 \\ L^{\alpha}_2 = c_1 - (c_1 - b_1)\alpha \end{cases} \qquad \text{and} \qquad \begin{cases} S^{\alpha}_1 = (b_2 - a_2)\alpha + a_2 \\ S^{\alpha}_2 = c_2 - (c_2 - b_2)\alpha \end{cases}$$

(III.6)

By multiplication operations,

$$R_1^{\alpha} = L^{\alpha}_1 \times S^{\alpha}_1 = (b_1 - a_1)(b_2 - a_2)\alpha^2 + [(b_1 - a_1)a_2 + (b_2 - a_2)a_1]\alpha + a_1 a_2$$

(III.7)

215

$$R_2^\alpha = L_2^\alpha \times S_2^\alpha = (c_1 - b_1)(c_2 - b_2)\alpha^2 + [(c_1 - b_1)c_2 + (c_2 - b_2)c_1]\alpha + c_1 c_2 \qquad \text{(III.8)}$$

For simplification,

If, $l_1 = (b_1 - a_1)(b_2 - a_2)$; $m_1 = (b_1 - a_1)a_2 + (b_2 - a_2)a_1$ and $u_1 = a_1 a_2$

$l_2 = (c_1 - b_1)(c_2 - b_2)$; $m_2 = (c_1 - b_1)b_2 + (c_2 - b_2)b_1$ and $u_2 = c_1 c_2$, then the solution will

be,

$$\alpha = \mu_R(x) = \begin{cases} \dfrac{-m_1 + \sqrt{(m_1^2 - 4l_1(u_1 - x))}}{2l_1} & a_1 a_2 < x < b_1 b_2 \\[3mm] \dfrac{m_2 + \sqrt{(m_2^2 - 4l_2(u_2 - x))}}{2l_2} & b_1 b_2 < x < c_1 c_2 \\[3mm] 0 & otherwise \end{cases}$$

$$\text{(III.9)}$$

Multiplication of vulnerability (crisp number) with the Equation II.26 will result the risk result in a fuzzy form. It should be noted that it would result the min and max operation and corresponding results.

Fuzzy Assessment Aggregation for consultation with several experts

This research uses the triangular fuzzy and trapezoidal shaped membership functions. The minimum, likely and maximum values of the membership functions can be selected after the consultation with the experts. However, consulted results likely have presence of bias and subjectivity. The bias and subjectivity of the expert view can be reduced by fuzzy average operation-which is also known as the "Triangular Average Formula". Further discussion on fuzzy operation is available in (Ngai and Wat, 2005). Let n be numbers evaluators are included in consultation. If they are assigning the values for the triangular fuzzy membership functions with a_1, a_2 and a_3 sides for fuzzy membership function A then the following relation are established.

$$A_i = (a_1^i, a_2^i, a_3^i), \qquad i = 1, \dots, n.$$

$$A_{average} = \frac{A_1 + A_2 + \dots + A_n}{n}, \quad A_{average} = \frac{(a_1^1, a_M^1, a_2^1) + \dots + (a_1^n, a_M^n, a_2^n)}{n} = \frac{\left(\sum_{i=1}^n a_1^i, \sum_{i=1}^n a_M^i, \sum_{i=1}^n a_2^i \right)}{n}$$

$$= \left(\frac{1}{n} \sum_{i=1}^n a_1^i, \frac{1}{n} \sum_{i=1}^n a_M^i, \frac{1}{n} \sum_{i=1}^n a_2^i \right) \qquad \text{(III.10)}$$

REFERENCES

Abebe, A.J. (2004). *Information theory and artificial intelligence to manage uncertainty in hydrodynamic and hydrological models*. PhD Thesis, Technical University, Delft, The Netherlands.

Abebe, A.J., Guinot, V. and Solomatine, D.P. (2000). Fuzzy alpha-cut vs. Monte Carlo techniques in assessing uncertainty in model parameters, Proc. of 4-th International Conference on Hydroinformatics, Iowa City.

Abrishamchi, A., Ebrahimian, A., Tajrishi, M. and Mariño, M.A. (2005). Case study: application of multicriteria cecision making to urban water supply. *Journal of Water Resources Planning and Management*, ASCE, 131(4), pp. 326-335

Adhikari, S. (2005). Asymptotic distribution method for structural reliability analysis in high dimensions. Proceedings of the Royal Society A: Mathematical, Physical and Engineering Science, 461(2062), pp. 3141-3158.

Agarwal, H., Renaud, J.E., Evan L. Preston and Padmanabhan, D. (2004). Uncertainty quantification using evidence theory in multidisciplinary design optimization. *Reliability Engineering & System Safety*, 85(1-3), pp. 281–294.

Ahmed, A., Kayis, B. and Amornsawadwatana, S. (2007). A review of techniques for risk management in projects. *Benchmarking: An International Journal*, 14(1), pp. 22 - 36.

Alcamo, J., Flörke, M. and Märker, M. (2007). Future long-term changes in global water resources driven by socio-economic and climatic changes. *Hydrological Sciences Journal,* 52(2), pp. 247-275.

Alcamo, J. and Kreileman, E. (1996). Emission scenarios and global climate protection. *Global Environmental Change*, 6(4), pp. 305-334.

Alegre, H., Baptista, J.M., Jr., E.C., Cubillo, F., Duarte, P., Hirner, W. and Wolf Merkel, R.P. (2006). Performance Indicators for Water Supply Services. IWA Publishing.

Ali, T. and Dutta, P. (2012). Modeling of Uncertainty in Dose Assessment using Probability-Possibility Transformation. *International Journal of Computer Applications,* 48(12), pp. 1-7.

Allen, M.R., Stott, P.A., Mitchell, J.F.B., Schnur, R. and Delworth, T.L. (2000). Quantifying the uncertainty in forecasts of anthropogenic climate change. *Nature*, 407, pp. 617-620.

Altunkaynak, A., Özger, M. and Çakmakci, M. (2005). Water consumption prediction of Istanbul city by using fuzzy logic approach. *Water Resources Management*, 19(5), pp. 641-654.

American Society of Civil Engineering (ASCE) (2005). The New Orleans hurricane protection system: What went wrong and why. A Report by the American Society of Civil Engineers Hurricane Katrina External Review Panel.

Ana, E. and Bauwens, W. (2010). Modeling the structural deterioration of urban drainage pipes: the state-of-the-art in statistical methods. *Urban Water Journal*, 7(1), pp. 47-59.

Ananda, J. and Herath, G. (2009). A critical review of multi-criteria decision making methods with special reference to forest management and planning. *Ecological Economics*, 68 pp. 2535–2548.

Ang, A.-H.S. and Tang, W.H. (1984). Probability concepts in engineering planning and design, Volume II Design decision, risk and reliability. John Wiley & Sons NY.

Ang, A.H.-S. and Tang, W.H. (1975). Probability Concepts in Engineering and Design Volume I-Basic Principles. John Wiley & Sons, Inc, New York, USA.

Anoop, M.B., Rao, K.B. and Gopalakrishnan, S. (2006). Conversion of probabilistic information into fuzzy sets for engineering decision analysis. *Computers and Structures*, 84 pp. 141–155.

Apel, H., Aronica, G., Kreibich, H. and Thieken, A. (2009a). Flood risk analyses—how detailed do we need to be? *Natural Hazards*, 49(1), pp. 79-98.

Apostolakis, G. (1990). The concept of probability in safety assessments of technological systems. *Science*, 250(4986), pp. 1359 - 1364.

Arnell, N.W. (2004). Climate change and global water resources: SRES emissions and socio-economic scenarios. *Global Environmental Change*, 14, pp. 31-52.

Arnell, N.W. and Delaney, E.K. (2006). Adapting to climate change: public water supply in England and Wales. *Climate Change, SpringerLink*, 78(2-4), pp. 227-255.

Ashley, R., Blackwood, D., Butler, D. and Jowitt, P. (2004). Sustainable Water Services: A Procedural Guide. Sustainable Water Industry Asset Resources Decisions Project. London: IWA Publishing.

Ashley, R., Blanksby, J., Cashman, A., Jack, L., Wright, G., Packman, J., Fewtrell, L., Poole, T. and Maksimovic, C. (2007). Adaptable urban drainage: Addressing change in intensity, occurrence and uncertainty of stormwater (AUDACIOUS). *Built Environment*, 33(1), pp. 70-84.

Aven, T. (2010). On the ned for restricting the probabilistic analysis in risk assessments to variability. *Risk Analysis*, 30(3), pp. 354-360.

Avena, T. and Zio, E. (2010). Some considerations on the treatment of uncertainties in risk assessment for practical decision making *Reliability Engineering & System Safety*, doi:10.1016/j.ress.2010.06.001.

Awasthi, A., Chauhan, S.S. and Goyal, S. (2010). A fuzzy multicriteria approach for evaluating environmental performance of suppliers. *International Journal of Production Economics*, 126(2), pp. 370-378.

Ayyub, B.M. and Chao, R.-J. (1998). Uncertainty Modelling in Civil Ebgineering. In: B.M. Ayub (ed.), Uncertainty Modeling and Analysis in Civil Engineering CRC Press, , New York.

Ayyub, B.M., McGill, W.L. and Kaminskiy, M. (2007). Critical Asset and Portfolio Risk Analysis: An All-Hazards Framework. *Risk analysis*, 27(4), pp. 789-801.

Babayan, A., Kapelan, Z., Savic, D. and Walters, G. (2005). Least-cost design of water distribution networks under demand uncertainty. *Journal of Water Resources Planning and Management*, 131(5), pp. 375-382.

Bae, H.-R., Grandhi, R.V. and Canfield, R.A. (2004). An approximation approach for uncertainty quantification using evidence theory. *Reliability Engineering & System Safety*, 86(3), pp. 215–225

Baraldi, P. and Zio, E. (2008). A combined Monte Carlo and possibilistic approach to uncertainty propagation in event tree analysis. *Risk Analysis*, 28(5), pp. 1309-1325.

Baudrit, C. and Dubois, D. (2005). Comparing methods for joint objective and subjective uncertainty propagation with an example in a risk assessment, 4th International Symposium on Imprecise Probabilities and Their Applications, Pittsburgh, Pennsylvania.

Baudrit, C., Dubois, D. and Guyonnet, D. (2006). Joint propagation and exploitation of probabilistic and possibilistic information in risk assessment. *IEEE Transactions On Fuzzy Systems*, 14(5), pp. 593-608.

Baudrit, C., Guyonnet, D. and Dubois, D. (2007). Joint propagation of variability and imprecision in assessing the risk of groundwater contamination. *Journal of Contaminant Hydrology*, 93 pp. 72–84.

Behzadian, M., Khanmohammadi Otaghsara, S., Yazdani, M. and Ignatius, J. (2012). A State-of-the-art Survey of TOPSIS Applications. *Expert Systems with Applications*, 39(17), pp. 13051–13069.

Belton, V. and Stewart, T. (2002). Multiple criteria decision analysis: an integrated approach. Kluwer Academic Publications, Boston.

Benayoun, R., De Montgolfier, J., Tergny, J. and Laritchev, O. (1971). Linear programming with multiple objective functions: Step method (STEM). *Mathematical programming*, 1(1), pp. 366-375.

Beven, K. and Binley, A. (1992). The future of distributed models: model calibration and uncertainty prediction. *Hydrological Processes*, 6(3), pp. 279-298.

Beven, K.J. and Alcock, R.E. (2011). Modelling everything everywhere: a new approach to decision-making for water management under uncertainty. *Freshwater Biology*, 57 (Suppl. 1), pp. 124–132.

Beven, K.J. and Alcock, R.E. (2012). Modelling everything everywhere: a new approach to decision-making for water management under uncertainty. *Freshwater Biology*, 57(s1), pp. 124-132.

Billings, R.B. and Agthe, D.E. (1998). State-space versus multiple regression for forecasting urban water demand. *Journal of Water Resources Planning and Management*, 124(2), pp. 113-117.

Bilgic, Taner & Turksen, Burhan. (1999) Measurement of Membership Function: Theoretical and Experimental Work. In D. Dubois and H. Prade (editors) Handbook of Fuzzy Systems, Vol. 1, Fundamentals of Fuzzy Sets, Chapter 3, pages 195-202. Kluwer Academic Publishers.

Biswas, A., K. (1991). Water for dustainable development in the 21st century: A global perspective. *Geo-journal*, 24(4), pp. 341-345.

Borja, Á., Galparsoro, I., Solaun, O., Muxika, I., Tello, E.M., Uriarte, A. and Valencia, V. (2006). The European Water Framework Directive and the DPSIR, a methodological approach to assess the risk of failing to achieve good ecological status. Estuarine, *Coastal and Shelf Science*, 66, pp. 84-96.

Bougadis, J., Adamowski, K. and Diduch, R. (2005). Short-term municipal water demand forecasting. *Hydrological Processes*, 19(1), pp. 137-148.

Bouwer, H. (2002). Integrated water management for the 21st century: problems and solutions. *Journal of Irrigation and Drainage*, 128(4), pp. 193-202.

Bradley, R.M. (2004). Forecasting domestic water use in rapidly urbanizing areas in Asia. Journal of *Environmental Engineering*, 130(4), pp. 465-471.

Brans, J.P., Vincke, P. and Mareschal, B. (1986). How to select and how to rank projects: The PROMETHEE method. *European Journal of Operational Research*, 24(2), pp. 228-238.

Brockwell, P.J. and Davis, R.A. (2002). Introduction to time series and forecasting. Springer Verlag.

Bucci, P., Kirschenbaum, J., Mangan, L.A., Aldemir, T., Smith, C. and Wood, T. (2008). Construction of event-tree/fault-tree models from a Markov approach to dynamic system reliability. *Reliability Engineering and System Safety*, 93, pp. 1616–1627.

Burr, D. (1994). A comparison of certain bootstrap confidence intervals in the Cox model. *Journal of the American Statistical Association*, 89(428), pp. 1290-1302.

Butler, D. and Maksimović, Ĉ. (2001). Interaction with the environment. In: Ĉ. Maksimović and J.A. Tejada-Guibert (eds.), Frontiers in Urban Water Management Deadlock or Hope, IWA Publishing, London, UK, pp. 84-142.

Cacciabue, P.C. (2005). Human error risk management methodology for safety audit of a large railway organisation. *Applied Ergonomics*, 36, pp. 709–718.

Carter, T.R., Hulme, M. and Viner, D. (1999). Representing Uncertainty in Climate Change Scenarios and Impact Studies. A Concerted Action Towards the Improved Understanding and Application of Results from Climate Model Experiments in European Climate Change Impacts Research, Climatic Research Unit, UEA, Norwich, UK.

Chen, Y., Hipel, K.W. and Kilgour, D.M. (2007). Multiple-criteria sorting using case-based distance models with an application in water resources management. *Systems, Man and Cybernetics, Part A: Systems and Humans, IEEE Transactions*, 37(5), pp. 680-691.

Chen, Y., Marc Kilgour, D. and Hipel, K.W. (2008). A case-based distance method for screening in multiple-criteria decision aid. *Omega*, 36(3), pp. 373-383.

Cheng, C. and Chau, K.W. (2001). Fuzzy iteration methodology for reservoir flood control operation1. *JAWRA Journal of the American Water Resources Association*, 37(5), pp. 1381-1388.

Chiu, S.L. (1994). Fuzzy model identification based on cluster estimation. *Journal of intelligent and Fuzzy systems*, 2(3), pp. 267-278.

Christensen, F.M., Andersen, O., Duijm, N.J. and Harremoës, P. (2003). Risk terminology-a platform for common understanding and better communication. *Journal of Hazardous Materials,Elsevier* 103(3), pp. 181-203.

Clarke, G.P., Kashti, A., McDonald, A. and Williamson, P. (1997). Estimating Small Area Demand for Water: A New Methodology. *Water and Environment Journal*, 11 (3), pp. 186 - 192.

Colyvan, M. (2008). Is probability the only coherent approach to uncertainty? *Risk Analysis*, 28(3), pp. 645-652.

Cooper, J.A., Ferson, S. and Ginzburg, L. (1996). Hybrid processing of stochastic and subjective uncertainty Data. *Risk Analysis*, 16(6), pp. 785–791.

Covello, V.T. and Mumpower, J. (1985). Risk Analysis and Management: Historical Perspectives. *Risk Analysis*, 5(2), pp. 103-120.

Cowell, S.J., Fairman, R. and Lofstedt, R.E. (2002). Use of risk assessment and life cycle Assessment in decision making: a common policy research agenda. Risk Analysis, 22(5), pp. 879-894.

Crouch, E.A.C. and Wilson, R. (1982). Risk/benefit analysis. Harper & Row, Publishers, Inc., Massachusetts.

Dasgupta, S. and Tam, E.K.L. (2005). Indicators and framework for assessing sustainable infrastructure. *Canadian Journal of Civil Engineering*, 32 pp. 30–44

Day Water (2003). Whitepaper on risk perception, risk assessment and risk management in the DayWater context, Project under EU RTD 5th Framework Programme.

De Marchi, B., Funtowicz, S., Lo Cascio, S. and Munda, G. (2000). Combining participative and institutional approaches with multicriteria evaluation. An empirical study for water issues in Troina, Sicily. *Ecological Economics*, 34(2), pp. 267-282.

Delgado, M. and Moral, S. (1987). On the concept of possibility-probability consistency Fuzzy Sets and Systems, 21(3), pp. 311-318

Delpla, I., Jung, A.-V., Baures, E., Clement, M. and Thomas, O. (2009). Impacts of climate change on surface water quality in relation to drinking water production. Environment International, 35, pp. 1225–1233.

Der Kiureghian, A. (1989). Measures of structural safety under imperfect states of knowledge. *Journal of Structural Engineering*, 115(5), pp. 1119-1140.

Deutsch, J.L. and Deutsch, C.V. (2012). Latin hypercube sampling with multi dimensional uniformity. *Journal of Statistical Planningand Inference*, 142, pp. 763–772.

Dhar, M. (2012). A note on the Coherence between Probability and Possibility Measures. *International Journal of Computer Applications,* 43(7).

Ding, K., Zhou, Z. and Liu, C. (1998). Latin hypercube sampling used in the calculation of the fracture probability. *Reliabililiy Engineering and System Safety,* 59, pp. 239-242.

Dubois, D. (2006). Possibility theory and statistical reasoning. *Computational Statistics & Data Analysis*, 51, pp. 47–69.

Dubois, D. (2010). Representation, propagation, and decision issues in risk analysis under incomplete probabilistic information. *Risk Analysis*, 30(3), pp. 361-368.

Dubois, D., Foulloy, L., Mauris, G. and Prade, H. (2004). Probability-possibility transformations, triangular fuzzy sets, and probabilistic inequalities. *Reliable Computing* 10, pp. 273–297.

Dubois, D. and Guyonnet, D. (2011). Risk-informed decision-making in the presence of epistemic uncertainty. *International Journal of General Systems*, 40(2), pp. 145-167.

Dubois, D. and Prade, H. (1980). Fuzzy sets and system-theory and applications. Academic Press, San Diego, California.

Dubois, D. and Prade, H. (1982). On several representations of uncertain body of evidence. In: M.M. Gupta and E.Sacnchez (eds.), Fuzzy information and decision making, North Holland Publishing, Amsterdam, pp. 167-181.

Dubois, D. and Prade, H. (1988). Possibility theory. Plenum Press, New York.

Dubois, D. and Prade, H. (1993). Fuzzy sets and probability: misunderstandings, bridges and gaps Proc. of Fuzzy Systems, 1993., Second IEEE International Conference on, San Francisco, CA, pp. 1059.

Dubois, D. and Prade, H. (1998). An introduction to fuzzy systems. *Clinica Chimica Acta*, 270, pp. 3–29.

Dubois, D., Prade, H. and Smets, P. (1996). Representing partial ignorance. *IEEE Transactions On Systems, Man, And Cybernetics-Part A: Systems And Humans*, 26(3).

Dubois, D., Prade, H. and Smets, P. (2001). New Semantics for Quantitative Possibility Theory. Lecture Notes in Computer Science, 2143, pp. 410-421.

Duijne, F.H.v., Aken, D.v. and Schouten, E.G. (2008). Considerations in developing complete and quantified methods for risk assessment. *Safety Science,* 46, pp. 245–254.

Durbach, I. and Stewart, T.J. (2003). Integrating scenario planning and goal programming. *Journal of Multi-Criteria Decision Analysis*, 12(4-5), pp. 261-271.

Eckart, J., Ghebremichael, K., Khatri, K., Tsegaye, S. and Vairavamoorthy, K. (2012). *Integrated Urban Water management for Mbale*, Uganda, The World Bank.

Efron, B. and R. J. Tibshirani (1993). *An Introduction to the Bootstrap.* Chapman and Hall, New York.

Elimelech, M. (2006). The global challenge for adequate and safe water. *Journal of Water Supply: Research and Technology—AQUA*, 55(1), pp. 3-10.

European Environment Agency (1999). Groundwater quality and quantity in Europe, Office for Official Publications of the European Communities, Luxembourg. Available at: http://reports.eea.europa.eu/groundwater07012000/en/enviassrp199903.

Ezell, B.C. (2007). Infrastructure vulnerability assessment model (I-VAM). *Risk Analysis*, 27(3), pp. 571-583.

Ezell, B.C., Bennett, S.P., Winterfeldt, D.v., Sokolowski, J. and Collins1, A.J. (2010). Probabilistic Risk Analysis and Terrorism Risk. *Risk Analysis*, 30(4), pp. 575-589.

Ezell, B.C., Farr, J.V. and Wiese, I. (2000). Infrastructure risk analysis of municipal water distribution system. *Journal of Infrastructure Systems*, 6(3), pp. 118-122.

Faber, M.H. and Stewart, M.G. (2003). Risk assessment for civil engineering facilities: critical overview and discussion. *Reliability Engineering and System Safety*, 80 pp. 173-184.

Farrelly, M. and Brown, R. (2011). Rethinking urban water management: Experimentation as a way forward? *Global Environmental Change*, 21(2), pp. 721-732.

Feng, G. (2006). A Survey on Analysis and Design of Model-Based Fuzzy Control Systems F*uzzy Systems, IEEE Transactions*, 14(5), pp. 676 - 697

Ferdous, R., Khan, F., Sadiq, R., Amyotte, P. and Veitch, B. (2011). Fault and event tree analyses for process systems risk analysis: uncertainty handling formulations. *Risk analysis*, 31(1), pp. 86-107.

Ferson, S. and Ginzburg, L.R. (1996). Different methods are needed to propagate ignorance and variability *Reliability Engineering & System Safety*, 54(2-3), pp. 133-144.

Ferson, S., Joslyn, C.A., Helton, J.C., Oberkampf, W.L. and Sentz, K. (2004). Summary from the epistemic uncertainty workshop: consensus amid diversity. *Reliability Engineering & System Safety*, 85(1-3), pp. 355-369.

Fisher, B. (2003). Fuzzy environmental decision-making: applications to air pollution. *Atmospheric Environment*, 37, pp. 1865–1877.

Flintsch, G.W. and Chen, C. (2004). Soft computing applications in infrastructure management. *Journal of Infrastructure Systems*, 10(4), pp. 157-166.

Florea, M.C., Jousselme, A.-L., Grenier, D. and Bosséc, É. (2008). Approximation techniques for the transformation of fuzzy sets into random sets. *Fuzzy Sets and Systems,* 159 ((2), pp. 270 – 288.

Geer, J.F. and Klir, G.J. (1992). A mathematical analysis of information-preserving transformations between probabilistic and possibilistic formulations of uncertainty. *International Journal of General Systems*, 20(2), pp. 143-176.

Geerse, J.M.U. and Lobbrecht, A.H. (2002). Assessing the performance of urban drainage systems: 'general approach' applied to the city of Rotterdam. *Urban Water* 4, pp. 199–209.

Gilovich, T., Griffin, D. and Kahneman, D. (2002). *Heuristics and Biases.* Cambridge University Press, New York.

Gober, P., Kirkwood, C.W., Balling, R.C., Ellis, A.W. and Deitrick, S. (2010). Water planning under climatic uncertainty in Phoenix: Why we need a new paradigm. *Annals of the Association of American Geographers*, 100(2), pp. 356-372.

Graedel, T.E. and Allenby, B.R. (2010). *Industrial ecology and sustainable engineering.* Prentice Hall Upper Saddle River, NJ, USA.

Guitouni, A. and Martel, J.-M. (1998). Tentative guidelines to help choosing an appropriate MCDA method. European Journal of Operational Research, 109 pp. 501-521.

Gupta, C.P. (1993). A note on the transformation of possibilistic information into probabilistic information for investment decisions. Fuzzy Sets and Systems 56, pp. 175-182.

Guyonnet, D., Bourgine, B., Dubois, D., Fargier, H., Côme, B. and Chilès, J.-P. (2003). Hybrid approach for addressing uncertainty in risk assessments. *Journal of Environmental Engineering*, 129(1), pp. 68-78.

Haimes, Y.Y. (1981). Hierarchical holographic modeling. *IEEE Transactions on System, Man and Cybernetics*, 11(9), pp. 606-617.

Haimes, Y.Y. (2009a). On the definition of resilience in systems. *Risk Analysis*, 29(4), pp. 498-501.

Haimes, Y.Y. (2009b). *Risk modeling, assessment, and management.* Wiley.

Haimes, Y.Y. and Leach, M.R. (1984). Decision making with multiobjectives. In: Y.Y. Haimes (ed.), *Risk Assessment and Management in a Multiobjective Framework*, pp. 23-25.

Haimes, Y.Y., Matalas, N.C., B.A.Jackson and James, F.R. (1998). Reducing vulnerability of water supply systems to attack. *Journal of Infrastructure Systems*, 4(4), pp. 164-177.

Haiyan, W. (2002). Assessment and prediction of overall environmental quality of Zhuzhou City, Hunan Province, China. *Journal of Environmental Management*, 66, pp. 329-340.

Hajkowicz, S. and Collins, K. (2007). A review of multiple criteria analysis for water resource planning and management. *Water Resources Management*, 21(9), pp. 1553-1566.

Hajkowicz, S. and Higgins, A. (2008). A comparison of multiple criteria analysis techniques for water resource management. *European Journal of Operational Research*, 184(1), pp. 255-265.

Hajkowicz, S., Spencer, R., Higgins, A. and Marinoni, O. (2008). Evaluating water quality investments using cost utility analysis. *Journal of Environmental Management*, 88(4), pp. 1601-1610.

Haldar, A. and Mahadevan, S. (2000). *Probability, reliability and statistical methods in engineering design.* John Wiley & sons, Inc., New York.

Hall, J. and Solomatine, D. (2008). A framework for uncertainty analysis in flood risk management decisions. *Intl. J. River Basin Management*, 6(2), pp. 85–98.

Hall, J., Watts, G., Keil, M., de Vial, L., Street, R., Conlan, K., O'Connell, P., Beven, K. and Kilsby, C. (2012). Towards risk-based water resources planning in England and Wales under a changing climate. *Water and Environment Journal*, 26(1), pp. 118-129.

Hall, J.W. (2003). Handling uncertainty in the hydroinformatic process. *Journal of Hydroinformatics*, 5(4), pp. 215-232.

Hämäläinen, R.P. and Alaja, S. (2008). The threat of weighting biases in environmental decision analysis. *Ecological Economics*, 68, pp. 556-569.

Hart, B., Burgman, M., Webb, A., Allison, G., Chapman, M., Duivenvoorden, L., Feehan, P., Grace, M., Lund, M., Pollino, C., Carey, J. and McCrea, A. (2005). Ecological risk management framework for the irrigation industry, Report to National Program for Sustainable Irrigation (NPSI) by Water Studies Centre, Monash University, Clayton, Australia. Available at: www.wsc.monash.edu.au.

Hasson, F., Keeney, S. and McKenna, H. (2000). Research guidelines for the Delphi survey technique. *Journal of advanced nursing*, 32(4), pp. 1008-1015.

Hellmuth, M. and Kabat, P. (2002). Impacts. In: B. Appleton (ed.), *Climate changes the water rules: How water managers can cope with today's climate variability and tomorrow's climate change*, Dialogue on Water and Climate, Delft.

Helton, J.C. and Davis, F.J. (2003). Latin hypercube sampling and the propagation of uncertainty in analyses of complex systems. *Reliability Engineering & System Safety*, 81(1), pp. 23-69.

Helton, J.C., Johnson, J.D. and Oberkampf, W.L. (2004). An exploration of alternative approaches to the representation of uncertainty in model predictions. *Reliability Engineering & System Safety*, 85(1), pp. 39-71.

Helton, J.C., Johnson, J.D., Oberkampf, W.L. and Sallaberry, C.J. (2010). Representation of analysis results involving aleatory and epistemic uncertainty. *International Journal of General Systems*, 39(6), pp. 605-646.

Helton, J.C., Johnson, J.D., Sallaberry, C.J. and Storlie, C.B. (2003). Survey of sampling-based methods for uncertainty and sensitivity analysi. *Reliability Engineering & System Safety*, 91(10–11), pp. 1175–1209.

Helton, J.C., Johnson, J.D., Sallaberry, C.J. and Storlie, C.B. (2006). Survey of sampling-based methods for uncertainty and sensitivity analysis. *Reliability Engineering & System Safety*, 91(10), pp. 1175-1209.

Ho, W. (2008). Integrated analytic hierarchy process and its applications-A literature review. *European Journal of Operational Research*, 186(1), pp. 211-228.

Hostmann, M., Bernauer, T., Mosler, H.J., Reichert, P. and Truffer, B. (2005). Multi-attribute value theory as a framework for conflict resolution in river rehabilitation. *Journal of Multi-Criteria Decision Analysis*, 13(2-3), pp. 91-102.

Huang, H.-W., Shih, C., Yih, S. and Chen, M.-H. (2008). System-level hazard analysis using the sequence-tree method. *Annals of Nuclear Energy* 35, pp. 353–362.

Hultman, N.E., Hassenzah, D.M. and Rayner, S. (2010). Climate Risk. *Environment and Resources*, 35 pp. 283-303.

Intergovernmental Panel on Climate Change (IPCC) (2007a). Working Group I: The Physical Science Basis, Regional Climate Projections-Africa, the Fourth Assessment Report of the Intergovernmental Panel on Climate Change Available at: http://ipcc-wg1.ucar.edu/wg1/wg1-report.html.

Intergovernmental Panel on Climate Change (IPCC) (2007b). Working Group II: Impacts, Adaptation and Vulnerability,Summary for Policymakers. Climate Change 2007 The Fourth Assessment Report of the Intergovernmental Panel on Climate Change Available at: http://www.ipcc.ch/SPM13apr07.pdf.

IPCC (2008). Climate change and water. the IPCC Working Group II.

IPCC (2012). Managing the Risks of Extreme Events and Disasters to Advance Climate Change Adaptation. Special Report of the Intergovernmental Panel on Climate Change, Cambridge University Press.

Jacobsen, M., Webster, M. and Vairavamoorthy, K. (2012). *The Future of Water in African Cities: Why Waste Water?* World Bank Publications.

Jacobi, S.K. and Hobbs, B.F. (2005). Quantifying and Mitigating the Splitting Bias and Other Value Tree-Induced Weighting Biases. *Decision Analysis*, 4(4), pp. 194-210.

Jain, A., Varshney, A.K. and Joshi, U.C. (2001). Short-term water demand forecast modelling at IIT Kanpur using artificial neural networks. *Water Resources Management*, 15(5), pp. 299-321.

Jang, J.S.R. (1993). ANFIS: Adaptive-network-based fuzzy inference system. *Systems, Man and Cybernetics, IEEE Transactions* on, 23(3), pp. 665-685.

Jasch, C. (2000). Environmental performance evaluation and indicators. *Journal of Cleaner Production*, 8, pp. 79–88.

Jha, A.K., Bloch, R. and Lamond, J. (2012). *Cities and Flooding, A Guide to Integrated Urban Flood Risk Management for the 21st Century.* The world Bank.

Jones, R.N., Dettmann, P., Park, G., Rogers, M. and White, T. (2007). The relationship between adaptation and mitigation in managing climate change risks: a regional response from North Central Victoria, Australia. *Mitigation and Adaptation Strategies for Global Change*, 12, pp. 685–712.

Jonkman, S.N. (2007). *Loss of Life Estimation in Flood Risk Assessment Theory and Applications.* PhD Thesis, Technical University Delft, Delft, The Netherlands.

Jumarie, G. (1994). Possibility-Probability Transformation: A New Result via Information Theory of Deterministic Functions. *Kybernetes*, 23 (5), pp. 56 - 59.

Kapelan, Z.S., Savic, D.A. and Walters, G.A. (2005). Multiobjective design of water distribution systems under uncertainty. *Water Resources Research*, 41 (W11407), pp. 1-15.

Kaplan, S. and Gerrick, J. (1981). "On the quantitative definition of risk". *Risk Analysis*, 1(1), pp. 11-27.

Kassab, M. (2006). *Integrated decision support system for infrastructure privatization under uncertainty using conflict resolution*, PhD Thesis, University of Waterloo.

Kaufmann, A. and Gupta, M.M. (1985). *Introduction to Fuzzy Arithmatic:Theory and Applications.* Van Nostrand, New York.

Keeney, R.L. (1996). *Value-focused thinking: A path to creative decisionmaking.* Harvard Univ Press.

Keeney, R.L. (2002). Common mistakes in making value trade-offs. *Operations Research*, 50(6), pp. 935–945.

Keeney, R.L. and Raiffa, H. (1993). *Decisions with multiple objectives: Preferences and value tradeoffs.* Cambridge Univ Pr.

Keilman, N., Pham, D.Q. and Hetland, A. (2002). Why population forecasts should be probabilistic - illustrated by the case of Norway. *Demographic Research*, 6(15), pp. 409-454.

Kelay, T., Chenoweth, J. and Fife-Schaw, C. (2006). Report on consumer trends cross-cutting issues across Europe, TECHNEAU. Available at: http://www.techneau.org/fileadmin/files/Publications/Publications/Deliverables/D1.1.12.pd.

Khadam, I.M. and Kaluarachchi, J.J. (2003). Multi-criteria decision analysis with probabilistic risk assessment for the management of contaminated ground water. *Environmental Impact Assessment Review*, 23(6), pp. 683-721.

Khatri, K. and Vairavamoorthy, K. (2007). Challenges for urban water supply and sanitation in the developing countries, Proc. of Water for changing world-Enhancing local knowledge and capacity, (13-15 June), UNESCO-IHE, Delft.

Khatri, K., Vairavamoorthy, K. and Porto, M. (2007). Challenges for urban water supply and sanitation in the developing countries. In: G.J. Alaerts and N.L. Dickinson (eds.), *Water for changing world, Developing local knowledge and capacity.*

Khatri, K.B. and Vairavamoorthy, K. (2009). Water demand forecasting for the city of the future against the uncertainties and the global change pressures: case of Birmingham,UK, Proc. of EWRI/ASCE: 2009, Kansas, USA, 2009 May (17-21), Kansas, USA

Khatri, K.B. and Vairavamoorthy, K. (2011a). A Hybrid Approach of Uncertainty Analysis for Performance Measurement of Water Distribution System, Proc. of, ASCE. 2011 April (11-13), Maryland, USA.

Khatri, K.B. and Vairavamoorthy, K. (2011b). A New Approach of Risk Analysis for Complex Infrastructure Systems under Future Uncertainties: A Case of Urban Water Systems, *Proc. of,* ASCE. 2011 April (11-13), Maryland, USA.

Khatri, K., and Vairavamoorthy, K (2011c). A new approach of decision making under risk and uncertainty while selecting a robust strategy. International conference on vulnerability and risk analysis and management (ICVRAM) and ISUMA 2011 fifth international symposium on uncertainty modelling and analysis by ASCE, 2011 April (11-13), Maryland, USA.

Khatri, K.B., Vairavamoorthy, K. and Akinyemi, E. (2011). A Framework for Computing a Performance Index for Urban Infrastructure Systems Using a Fuzzy Set Approach. *Journal of Infrastructure Systems*, 1, pp. 33.

Khatri, K., and Vairavamoorthy, K. (2013a). Major sources of risk and uncertainty urban water system due to future change pressures — a review. *Journal of Urban Water*, In submission.

Khatri, K., Vairavamoorthy, K. and Ali, Y. (2013b). A Hybrid approach to uncertainty analysis for a long-term planning of urban water system—A case of future water demand analysis in Birmingham, UK, *ASCE Journal of Water Resources Planning and Management*, In submission.

Khatri, K., and Vairavamoorthy, K. (2013c). Decision making under risk and uncertainty –a case of Mbale Urban Water System, Uganda. *ASCE Journal of Water Resources Planning and Management,* In submission.

Kleiner, Y. (1998). Risk factors in water distribution systems, 26th Annual Conference, British Columbia Water and Waste Association, Whistler, B.C., Canada.

Kleiner, Y. and Rajani, B. (2001). Comprehensive review of structural deterioration of water mains: statistical models.*Urban Water*, 3(3), pp. 131-150.

Kleiner, Y., Rajani, B. and Sadiq, R. (2006). Failure risk management of buried infrastructure using fuzzy-based techniques. *Journal of Water Supply Research and Technology: Aqua*, 55(2), pp. 81-94.

Kleiner, Y. (1998). Risk factors in water distribution systems,, 26th Annual Conference, British Columbia Water and Waste Association, Whistler, B.C., Canada.

Kleiner, Y., Sadiq, R. and Rajani, B. (2004). Modeling failure risk in buried pipes using fuzzy Markov deterioration process. NRCC-46739, pp. 1-13.

Klir, G. and Yuan, B. (1995). *Fuzzy sets and fuzzy logic theory and applications.* Prentice Hall, New Jersey.

Klir, G.J. (1987). Where do we stand on measures of uncertianty, amguity, fuzziness and the like? *Fuzzy Sets and Systems, Elsevier Science Publishers*, 24, pp. 141-160, very good.

Klir, G.J. (1990a). A principle of uncertainty and information invariance. *International Journal of General Systems*, 17(2), pp. 249-275.

Klir, G.J. (1990b). A principle of uncertainty and information invariance. *International Journal of General Systems*, 17(2), pp. 249 — 275.

Klir, G.J. and Parviz, B. (1992). Probability-possibility transformations: a comparison. International Journal of General Systems, 21(3), pp. 291 - 310Klir, G.J. (1995). Principles of uncertainty: What are they? Why do we need them? *Fuzzy Sets and Systems* 74, pp. 15- 31.

Klir, G.J. (2006). *Uncertainty and information: foundations of generalized information theory*. Wiley-Interscience, Hoboken, New Jersey.

Klir, G.J. and Folger, T.A. (1998). *Fuzzy sets, uncertainty and information*. Prentice Hall International, London.

Klir, G.J. and Smith, R.M. (2001). On measuring uncertainty and uncertainty-based information: Recent developments. *Annals of Mathematics and Artificial Intelligence* 32, pp. 5–33.

Klutke, G.-A., Kiessler, P.C. and Wortman, M. (2003). A critical look at the bathtub curve. *Reliability, IEEE Transactions on*, 52(1), pp. 125-129.

Knight, F.H. (1921). *Risk, Uncertainty, and Profit* Hart, Schaffner & Marx,, Boston, MA.

Kolsky, P. and Butler, D. (2002). Performance indicators for urban storm drainage in developing countries. *Urban Water* 4, pp. 137-144.

Kovats, S. and Akhtar, R. (2008). Climate, climate change and human health in Asian cities. *Environment and Urbanization* 20(1), pp. 165-175.

Kristensena, V., Avenb, T. and Ford, D. (2006). A new perspective on Renn and Klinke's approach to risk evaluation and management. *Reliability Engineering and System Safety*, 91, pp. 421-432.

Krzysztofowciz, R. (2001). The case for probabilistic forecasting in hydrology. *Journal of hydrology*, 249(1), pp. 2-9.

Kuczera, G. and Parent, E. (1998). Monte Carlo assessment of parameter uncertainty in conceptual catchment models: the Metropolis algorithm. *Journal of Hydrology*, 211(1), pp. 69-85.

Kunsch, H.R. (1989). The jackknife and the bootstrap for general stationary observations. *The Annals of Statistics*, 17(3), pp. 1217-1241.

Lee, R.D. and Tuljapurkar, S. (1994). Stochastic population forecasts for the United States: Beyond high, medium, and low. *Journal of the American Statistical Association*, 89, pp. 1175– 1189.

Limbourg, P. and Rocquigny, E.d. (2010). Uncertainty analysis using evidence theory – confronting level-1 and level-2 approaches with data availability and computational constraints. *Reliability Engineering & System Safety*, 95(5), pp. 550–564.

Lindhe, A., Rosén, L., Norberg, T. and Bergstedt, O. (2009). Fault tree analysis for integrated and probabilistic risk analysis of drinking water systems. *Water Research*, 43(6), pp. 1641-1653

Linkov, I., Satterstrom, F.K., Kike, G., Batchelor, C., Bridges, T. and Ferguson, E. (2006). From comparative risk assessment to multi-criteria decision analysis and adaptive management: Recent developments and applications. *Environment International* 32, pp. 1072–1093.

Liu, H.C., Liu, L. and Liu, N. (2012). Risk evaluation approaches in failure mode and effects analysis: A literature review. *Expert Systems with Applications*, 6, pp. 1195–1207.

Liu, R.Y. and Singh, K. (1992). Moving blocks jackknife and bootstrap capture weak dependence. *Exploring the limits of bootstrap*, 225, pp. 248.

Lu, R.-S. and Lo, S.-L. (2002). Diagnosing reservoir water quality using self-organizing maps and fuzzy theory. *Water Research*, 36(9), pp. 2265-2274.

Lu, R.S., Lo, S.L. and Hu, J.Y. (1999). Analysis of reservoir water quality using fuzzy synthetic evaluation. *Stochastic Environmental Research and Risk Assessment*, 13, pp. 327-336.

Macdonald, I. and Strachan, P. (2001). Practical application of uncertainty analysis. *Energy and Buildings* 33, pp. 219-227

Malmqvist, P.A., Heinicke, G., Karrman, E., Stenstrom, T.A. and Svensson, G. (eds.) (2006). *Strategic Planning of Sustainable Urban Water Management*. IWA Publishing, London.

Mannina, G., Freni, G., Viviani, G., Sægrov, S. and Hafskjold, L.S. (2006). Integrated urban water modelling with uncertainty analysis. *Water Science & Technology*, 54(6), pp. 379-386.

Marsalek, J., Jiménez-Cisneros, B.E., Malmquist, P.-A., Karamouz, M. and Chocat, J.G.a.B. (2006). Urban water cycle processes and interactions. In: I.S.U.D.o.W. Sciences and P.C. 1 rue Miollis, France (Editors). IHP-VI Technical Document in Hydrology , UNESCO Working Series SC-2006/WS/7.

Maskey, S. (2004). *Modelling Uncertainty in Flood Forecasting Systems*. PhD Thesis, Technical University Delft, Delft, The Netherlands.

Maskey, S., Guinot, V. and Price, R.K. (2004). Treatment of precipitation uncertainty in rainfall-runoff modelling: a fuzzy set approach. *Advances in Water Resources* 27, pp. 889–898.

Matos, R., Cardoso, A., Duarte, P., Ashley, R. and Molinari, A. (2003). *Performance indicators for wastewater services*. International Water Assn.

Mauris, G. (2010). Transformation of Bimodal Probability Distributions Into Possibility Distributions. *IEEE Transactions On Instrumentation And Measurement*, 59(1), pp. 39-47.

McDonald, R.I., Green, P., Balk, D., Fekete, B.M., Revenga, C., Todd, M. and Montgomery, M. (2011). Urban growth, climate change, and freshwater availability. *Proceedings of the National Academy of Sciences*, 108(15), pp. 6312-6317.

McGill, W.L., Ayyub, B.M. and Kaminskiy, M. (2007). Risk Analysis for Critical Asset Protection. *Risk analysis*, 27(5), pp. 1265-1281.

McIntyre, N.R., Wagener, T., Wheater, H.S. and Yu, Z.S. (2003). Uncertainty and risk in water quality modelling and management. *Journal of Hydroinformatics*, 5(4), pp. 259-274.

McIntyre, N.R. and Wheater, H.S. (2004). A tool for risk-based management of surface water quality. *Environmental Modelling & Software* 19, pp. 1131–1140.

McKay, M.D. (1992). Latin hypercube sampling as a tool in uncertainty analysis of computer models, *Proc. of Proceedings of the 24th conference on Winter simulation*, Arlington, Virginia,, pp. 557-564.

Mehrotra, S., Natenzon, C.E., Omojola, A., Folorunsho, R., Joseph Gibride, J. and Rosenzweig, C. (2009). *Framework for city climate risk assessment* World Bank Commissioned Research, Marseille, France.

Mendoza, G. and Martins, H. (2006). Multi-criteria decision analysis in natural resource management: A critical review of methods and new modelling paradigms. *Forest Ecology and Management*, 230(1), pp. 1-22.

Mendoza, G.A. and Prabhu, R. (2005). Combining participatory modeling and multi-criteria analysis for community-based forest management. *Forest Ecology and Management*, 207(1), pp. 145-156.

Mitchell, V., Mein, R.G. and McMahon, T.A. (2001). Modelling the urban water cycle. *Environmental Modelling & Software*, 16(7), pp. 615-629.

Mitchell, V. and Diaper, C. (2006). Simulating the urban water and contaminant cycle. Environmental Modelling & Software, 21(1), pp. 129-134.

Milman, A. and Short, A. (2008). Incorporatingresilienceintosustainabilityindicators:Anexampleforthe urban watersector. Global EnvironmentalChange, 18, pp. 758-767.

Misiunas, D. (2006). *Failure Monitoring and Asset Condition Assessment in Water Supply Systems*, Lund University, Lund, Sweden.

Montanari, A. and Brath, A. (2004). A stochastic approach for assessing the uncertainty of rainfall-runoff simulations. *Water Resources Research*, 40(1), pp. W01106.

Morgan, M. and Henrion, M. (1990). *A guide to dealing with uncertainty in quantitative risk and policy analysis*. Cambridge University Press.

Moss, R.H., Edmonds, J.A., Hibbard, K.A., Manning, M.R., Rose, S.K., van Vuuren, D.P., Carter, T.R., Emori, S., Kainuma, M. and Kram, T. (2010). The next generation of scenarios for climate change research and assessment. *Nature*, 463(7282), pp. 747-756.

Mukheibir, P. and Ziervogel, G. (2007). Developing a Municipal Adaptation Plan (MAP) for Climate Change: The City of Cap Town. *Environment and Urbanization*, 19(1), pp. 143-158.

Munda, G. (1995). *Multicriteria evaluation in a fuzzy environment: theory and applications in ecological economics*. Physica-Verlag.

Murray, A.T., Matisziw, T.C. and Grubesic, T.H. (2008). A methodological overview of network vulnerability analysis. *Growth and Change*, 39(4), pp. 573–592.

Mythen, G. and Walklate, S. (2008). Terrorism, Risk and International Security: The Perils of Asking 'What If?'. *Special Issue on Security, Technologies of Risk, and the Political*, 39(2-3), pp. 221–242.

Nakicenovic, N., Victor, N. and Morita, T. (1998). Emissions scenarios database and review of scenarios. *Mitigation and Adaptation Strategies for Global Change*, 3(2–4), pp. 95–131.

National Water and Sewerage Corporation (2012). Personal Communication with Mr Charles Okuonzi on Water Supply Situation in Mbale Town, Area Manager Mbale Office, NWSC, Uganda.

Nicola, A.D. and McCallister, A. (2006). Existing Experiences of Risk Assessment. *European Journal on Criminal Policy and Research*, 12(3-4), pp. 179-187, saved deom tudelft.

NRC (1983). *Risk assessment in the federal government: managing the process, committee for the institutional means for assessment of risks to public health*, National Academic Press, Washington D.C.

NRC (1996). Measuring and Improving Infrastructure Performance. In: Commission on Engineering and Technical System (Editor). National Research Centre (NRC), National Academies Press, Washington DC.

NRC (2000). *Beyond six billion: forecasting the World's population. Panel on populatoin projections.*, Committee on Populatoin, Commission on Behavioral and Social Sciences and Education, National Academy Press, Washington, DC. Available at: http://www.nap.edu.

Öberg, T. and Bergbäck, B. (2005). A Review of Probabilistic Risk Assessment of Contaminated Land. *J Soils & Sediments* 5(4), pp. 213 – 224.

Oberkampf, W.L., DeLand, S.M., Rutherford, B.M., Diegert, K.V. and Alvin, K.F. (2002). Error and uncertainty in modeling and simulation. *Reliability Engineering & System Safety*, 75(3), pp. 333-357.

Oussalah, M. (1999). On the probability/possibility transformations: A comparative analysis. *International Journal of General Systems*, 29(5), pp. 671-718.

Paneque Salgado, P., Corral Quintana, S., Guimaraes Pereira, A., del Moral Ituarte, L. and Pedregal Mateos, B. (2009). Participative multi-criteria analysis for the evaluation of water governance alternatives. A case in the Costa del Sol (Malaga). *Ecological Economics*, 68(4), pp. 990-1005.

Pappenberger, F. and Beven, K.J. (2006). Ignorance is bliss:Or seven reasons not to use uncertainty analysis. *Water Resources Research* 42(5), pp. 1-8.

Pappenberger, F., Harvey, H., Beven, K., Hall, J., Romanowicz, R. and Smith, P. (2005). *Risk & Uncertainty Tools and Implementation*, Civil Engineering and Geosciences, University of Newcastle, UK.

Parry, G.W. (1996). The characterization of uncertainty in probabilistic risk assessment of complex systems. 54(2), pp. 119-126.

Paté-Cornell, M.E. (1996). Uncertainties in risk analysis: six levels of treatment *Reliability Engineering and System Safety*, 54(2), pp. 95-111.

Pedrycz, W. (1993). *Fuzzy control and fuzzy systems* (2nd extended edition) Research Studies Press/J. Wiley, Taunton/New York

Pinkham, R. (1999). *21st century water systems: scenarios, visions, and drivers*, Rocky Mountain Institute, Snowmass, Colorado. Available at: http://www.rmi.org/images/other/Water/W99-21_21CentWaterSys.pdf.

Pohekar, S.D. and Ramachandran, M. (2004). Application of multi-criteria decision making tosustainable energy planning-A review. *Renewable and Sustainable Energy Reviews*, 8 pp. 365–381.

Price, M.F. (1989). Global change: Defining the ill-defined. *Environment*, 31(8), pp. 18-20, 42-44.

Prudhomme, C., R.L. Wilby, S. Crooks, A.L. Kay and Reynard, N.S. (2010). Scenario-neutral approach to climate change impact studies: Application to flood risk. *Journal of Hydrology*, 390(3–4), pp. 198–209.

Rajani, B., Kleiner, Y. and Sadiq, R. (2006). Translation of pipe inspection results into condition ratings using the fuzzy synthetic evaluation technique. *Journal of water supply and technology: Aqua*, 55(1), pp. 11-24.

Raje, D.V., Dhobe, P.S. and Deshpande, A.W. (2002). Consumer's willingness to pay more for municipal supplied water: a case study *Ecological Economics*, 42(3), pp. 391-400.

Ralescu, A.L. and Ralescu, D.A. (1984). Probability and fuzziness. *Information Sciences* 34, pp. 85-92.

Renn, O. (1998). Three decades of risk research: accomplishments and new challenges. *Journal of Risk Research*, 1(1), pp. 49-71.

Rijsbermana, M.A. and Van de Ven, F.H.M. (2000). Different approaches to assessment of design and management of sustainable urban water systems. *Environmental Impact Assessment Review*, 20 pp. 333–345.

Roccaro, P., Mancini, G. and Vagliasindi, F.G.A. (2005). Water intended for human consumption— Part I: Compliance with European water quality standards. *Desalination*, 176(1), pp. 1-11.

Romero, C. and Rehman, T. (1987). Natural resource management and the use of multiple criteria decision-making techniques: a review. *European Review of Agricultural Economics*, 14(1), pp. 61-89.

Ross, T.J. (2008). *Fuzzy logic with engineering applications*. Wiley Online Library.

Rotmans, J., Asslelt, M.B.A.v. and Vries, B.J.M.d. (1997). Global change and sustainable development In: J. Rotmans and B.d. Vries (eds.), *Perspectives on global change, the targets approach*, Cambridge University Press, Cambridge.

Rowe, W.D. (1977). *An Anatomy of Risk*. John Wiley & Sons, New York.

Roy, B. (2005). Paradigms and Challenges. In: J.Figueira, S. Greco and M. Ehrgott (eds.), *An overview of MCDA techniques today, multiple criteria decision analysis: state of the art surveys*, Springer Science, Boston.

Roy, B. and Vanderpooten, D. (1996). The European School of MCDA: emergence, basic features and current works. *Journal of Multi-Criteria Decision Analysis*, 5(1), pp. 22-38.

Ruth, M., Bernier, C., Jollands, N. and Golubiewski, N. (2007). Adaptation of urban water supply infrastructure to impacts from climate and socioeconomic changes: The case of Hamilton, New Zealand. *Water Resour Management*, 21, pp. 1031–1045.

Saaty, T.L. (1988). *Multicriteria decision-making: The analytic hierarchy process*. University of Pittsburgh, Pittsburgh.

Saaty, T.L. and Cho, Y. (2001). The decision by the US congress on China's trade status: a multicriteria analysis. *Socio-Economic Planning Sciences*, 35(4), pp. 243-252.

Sadeghi, N., Fayek, A.R. and Pedrycz, W. (2010). Fuzzy Monte Carlo Simulation and Risk Assessment in Construction. *Computer-Aided Civil and Infrastructure Engineering* 25 pp. 238–252.

Sadiq, R., Kleiner, Y. and Rajani, B. (2004). Aggregative risk analysis for water quality failure in distribution networks. *Journal of Water Supply Research and Technology : Aqua*, 53(4), pp. 241-261.

Sadiq, R., Kleiner, Y. and Rajani, B. (2007). Water quality failures in distribution networks – risk analysis using fuzzy logic and evidential reasoning. *Risk Analysis*, 27(5), pp. 1381-1394.

Sadiq, R. and Rodriguez, M.J. (2004). Fuzzy synthetic evaluation of disinfection by-products-a risk-based indexing system. *Journal of Environmental Management*, 73, pp. 1-13.

Sægrov, S., Baptista, J.F.M., Conroy, P., Herz, R.K., LeGau   re, B. and Schiatti, M. (1999). Rehabilitation of water networks survey of research needs and on-going efforts. *Urban Water*, 1, pp. 15-22.

Schiedek, D., Sundelin, B., Readman, J.W. and Macdonald, R.W. (2007). Interactions between climate change and contaminants. *Marine Pollution Bulletin*, 54, pp. 1845-1856.

Schneider, M.L. and Whitlatch, E.E. (1991). User-specific water demand elastcities. *Journal of Water Resources Planning and Management*, 117(1), pp. 52-73.

Segrave, A.J. (2007). *Report on trends in the Netherlands*, TECHNEAU.

Seica, M.V. and Packer, J.A. (2004). Mechanical properties and strength of aged cast iron water pipes. *Journal of Materials in Civil Engineering*, 16(1), pp. 69-77.

Sentz, K. and Ferson, S. (2002). *Combination of Evidence in Dempster-Shafer Theory*, Sandia National Laboratories, Binghamton, NY SAND 2002-0835

Serrurier, M. and Prade, H. (2011). Maximum-likelihood principle for possibility distributions viewed as families of probabilities *Proc. of 2011 IEEE International Conference on Fuzzy Systems*, Taipei, Taiwan, pp. 2987-2993.

Severn Trent Water Limited (STWL) (2007). *Gloucestershire 2007 the impact of the July Floods on the water infrastructure and customer service final report*

Severn Trent Water Limited (STWL) (2008). *Severn Trent Water - Water Resources Management Plan 2009.*

Shafer, G. (1976). *A Mathematical Theory of Evidence*. Princeton, NJ: Princeton University Press.

Shannon, C.E. (1948). A Mathematical Theory of Communication. *Bell System Technical Journal*, 27(379-423), pp. 623-56.

Shrestha, D.L. (2009). *Uncertainty Analysis in Rainfall -Runoff modelling: Application of Machine Learning Technques*. PhD Thesis, UNESCO-IHE/TU Delft, Delft.

Silvert, W. (2000). Fuzzy indices of environmental conditions *Ecological Modelling*, 130(1-3), pp. 111-119

Sipahi, S. and Timor, M. (2010). The analytic hierarchy process and analytic network process: an overview of applications. *Management Decision*, 48(5), pp. 775-808.

Slovic, P. (1987). Perception of Risk. *Science*, 236(4799), pp. 280-285.

Stan Kaplan, Yacov Y. Haimes and Garrick, B.J. (2001). Fitting hierarchical holographic modeling into the theory of scenario structuring and a resulting refinement to the quantitative definition of risk. *Risk Analysis*, 21(5), pp. 807-819.

Stein, M. (1987). Large sample properties of simulations using Latin hypercube sampling. *Technometrics*, 29(2), pp. 143-151.

Steuer, R.E. and Na, P. (2003). Multiple criteria decision making combined with finance: A categorized bibliographic study. *European Journal of Operational Research*, 150(3), pp. 496-515.

Stirling, A. (2006). Analysis, participation and power: justification and closure in participatory multi-criteria analysis. *Land Use Policy*, 23(1), pp. 95-107.

Susan, K.S., Eugene, A. and Shu-Chen, W. (1988). Latin Hypercube Sampling and the Sensitivity Aalysis of a Monte Carlo Epidemic Model. *Int J BiomedComput, Elsevier Scientific Publishers Ireland Ltd.*, 23, pp. 97-112.

SuterII, G.W. (2007). *Ecological Risk Assessment*. Taylor and Francis Group, New York, USA.

SWITCH (2006). *Managing Water for the City of the Future*. UNESCO-IHE.

Talvitie, A. (1999). Performance indicators for the road sector. *Transportation research part -A*, 26(1), pp. 5-30.

Tao, Y. and Xinmiao, Y. (1998). Fuzzy comprehensive assessment, fuzzy clustering analysis and its application for urban traffic environment quality evaluation. *Transportation Research Part D: Transport and Environment*, 3(1), pp. 51-57.

Tonon, F., Bernardini, A. and Elishakoff, I. (2001). Hybrid analysis of uncertainty: probability, fuzziness and anti-optimization. *Chaos, Solitons & Fractals*, 12(8), pp. 1403-1414.

Tung, Y.-K. (1996). Uncertainty and reliability analysis. In: L.W. Mays (ed.), *Water Resources Handbook* McGraw-Hill Book Company, pp. 7.1-7.65.

U.S. Environmental Protection Agency (EPA) (2003). *Framework for Cumulative Risk Assessment*, Risk Assessment Forum,U.S. Environmental Protection Agency EPA/630/P-02/001F, Washington DC.

U.S. Environmental Protection Agency (EPA) (2004). *An Examination of EPA Risk Assessment Principles and Practices*. EPA/100/B-04/001, Office of the Science Advisor U.S. Environmental Protection Agency, EPA/100/B-04/001Washington.

Uganda Bureau of Statistics (2011). Uganda Bureau of Statistics Official Site. http://www.ubos.org/?st=pagerelations2&id=17&p=related%20pages%202:Population.

Van Asselt, M.B.A. and Rotmans, J. (2002). Uncertainty in Integrated Assessment Modelling *Climatic Change*, 54(1/2), pp. 75-105.

Van der Klis, H. (2003). *Uncertainty analysis applied to numerical models of river bed morphology*. PhD Thesis Thesis, Technical University of Delft, Delft, The Netherlands.

Viana, F.A., Venter, G. and Balabanov, V. (2010). An algorithm for fast optimal Latin hypercube design of experiments. *International journal for numerical methods in engineering*, 82(2), pp. 135-156.

Vlek, C.A.J. (1996). A multi-level, multi-stage and multi-attribute perspective on risk assessment, decision-making and risk control *Risk Decision and Policy*, 1(1), pp. 9-31.

Vörösmarty, C.J., Green, P., Salisbury, J. and Lammers, R.B. (2000). Global Water Resources: Vulnerability from Climate Change and Population Growth. *Science*, 289, pp. 284-287.

Vörösmarty, C.J., McIntyre, P., Gessner, M.O., Dudgeon, D., Prusevich, A., Green, P., Glidden, S., Bunn, S.E., Sullivan, C.A. and Liermann, C.R. (2010). Global threats to human water security and river biodiversity. *Nature*, 467(7315), pp. 555-561.

Walley, P. (1991). *Statistical reasoning with imprecise probabilities*. Chapman and Hall, New York.

Wang, C.-h. and Blackmore, J.M. (2008). Resilience concepts for water resource systems. *Journal of Water Resources Planning and Management*, 135(6), pp. 528-536.

Wang, J.J., Jing, Y.Y., Zhang, C.F. and Zhao, J.H. (2009). Review on multi-criteria decision analysis aid in sustainable energy decision-making. *Renewable and Sustainable Energy Reviews*, doi:10.1016/j.rser.2009.06.021.

WCED (1987). *Our Common Future*. World Commission on Environment and Development (WCED). Oxford University Press, Oxford.

White, C., Victory, T.F. and Mulhern, L. (1999). Using hierarchical organization to reduce complexity in water distribution system simulation models, *Proc. of WRPMD 1999 — Preparing for the 21st Century, Proceedings of 29th Annual Water Resources Planning and Management Conference* ASCE.

WHO (2001). *Report on integrated risk assessment*, World Health Organisation, Geneva, Switzerland.

Wilby, R.L. and Dessai, S. (2010). Robust adaptation to climate change. *Weather*, 65(7), pp. 180-185.

Wilby, R.L., Troni, J., Biot, Y., Tedd, L., Hewitson, B.C., Smith, D.G. and Sutton, R.T. (2009). A review of climate risk information for adaptation and development planning. *International Journal of Climatology*, 29(9), pp. 1193–1215.

Willows, R.I. and Connell, R.K. (eds.) (2003). UKCIP Technical Report, Oxford.

Wilson, T. and Bell, M. (2004). Australia's uncertain demographic future. *Demograhpic Research*, 11(8), pp. 195-234

Yager, R.R. (1986). A characterization of the extension principle. *Fuzzy Sets and Systems*, 18(3), pp. 205-217.

Yager, R.R. (1987). On the dempster-shafer framework and new combination rules. *Information Sciences*, 41, pp. 93-137

Yamada, K. (2001). Probability -possibility transformation based on evidence theory, *Proc. of IFSA World Cong. and 20th NAFIPS International Conferenc* IEE Xplore, pp. 70-75.

Yan, J.M. and Vairavamoothy, K. (2003). Fuzzy approach for the pipe condition assessment, New Pipeline Technologies, Security, and Safety, *Proc. of Proceeding of ASCE international conference on pipeline engineering and construction*, Baltimore, Maryland, USA, American Society of Civil Engineers pp. 1817.

Yao, W., Chen, X., Luo, W., Tooren, M.v. and Guo, J. (2011). Review of uncertainty multidisciplinary design optimization methods for aerospace vehicles. *Porgress in Aerospace Sciences*, 47, pp. 450-479.

Yen, B. and Ang, A.S. (1971). Risks analysis in design of hydraulic projects. In: C.Chiu (ed.), *Proc. of Stochastic Hydraulics*, University of Pittsburg, Pittsburg, PA, USA, pp. 694-709.

Zadeh, L.A. (1965). Fuzzy sets. *Inf. Control*, 8, pp. 338–353.

Zadeh, L.A. (1975). The concept of a linguistic variable and its application to approximate reasoning—I. *Information Sciences*, 8(3), pp. 199-249.

Zadeh, L.A. (1978). Fuzzy sets as a basis for a theory of possibility. *Fuzzy Sets and Systems* 100(1), pp. 9-34.

Zadeh, L.A. (1983). The role of fuzzy logic in the management of uncertainty in expert systems. *Fuzzy Sets and Systems*, 11(1-3), pp. 197-198.

Zadeh, L.A. (2008). Is there a need for fuzzy logic? *Information Sciences*, 178(13), pp. 2751-2779.

Zahedi, F. (1986). A simulation study of estimation methods in the analytic hieracrchy process. *Scoio-Economical Planning Science*, 20(6), pp. 347-354.

Zedeh, L. (1989). Knowledge representation in fuzzy logic. Knowledge and Data Engineering, IEEE Transactions on, 1(1), pp. 89-100.

Zeleny, M. (1981). Multiple criteria decision making. McGraw-Hill, New York.

Zhao, Y.-G. and Ono, T. (2001). Moment methods for structural reliability Structural safety, 23, pp. 47-75.

Zhou, S.-L., McMahon, T.A., Walton, A. and Lewis, J. (2000). Forecasting daily urban water demand: a case study of Melbourne. *Journal of Hydrology*, 236(3), pp. 153-164.

Zimmermann, H.J. (1987). *Fuzzy sets, decision making, and expert systems*, 10. Springer.

Zopounidis, C. and Doumpos, M. (2002). Multi-criteria decision aid in financial decision making: methodologies and literature review. *Journal of Multi-Criteria Decision Analysis*, 11(4-5), pp. 167-186.

Zoubir, A.M. and Robert Iskander, D. (2007). Bootstrap methods and applications. *Signal Processing Magazine, IEEE*, 24(4), pp. 10-19.

SAMENVATTING

Toekomst op lange termijn planning is geen nieuwe aanpak in stedelijk waterbeheer (SWB). Echter, de conventionele 'stationaire benadering' van infrastructuur planning en besluitvorming, waarin de toekomst wordt beschouwd als de voortzetting van de historische waarnemingsperiode, zal niet werken in de huidige snel veranderende omgeving. Dit komt omdat in de huidige en toekomstige veranderingen klimaatverandering, verstedelijking, bevolkingsgroei, verslechtering van de infrastructuur systemen, en veranderingen in de sociaal-economische omstandigheden altijd onzeker zijn. Onzekerheid in veranderingen in de toekomst komt voort uit twee verschillende bronnen: onvolledige kennis en ontbrekende kennis. Onvolledige kennis is te wijten aan een gebrek aan informatie en begrip van een systeem. Ontbrekende kennis is te wijten aan de inherente onbepaaldheid van de toekomstige menselijke samenlevingen en natuurlijke en gebouwde systemen.

Stedelijke watersystemen (SWS) zijn complex en dynamisch. Ze bestaan uit een groot aantal op elkaar inwerkende componenten en meerdere beslissers: de gebruikers, waterbedrijven, gemeentelijke besturen, en andere autoriteiten. Dergelijke complexe interacties en onderlinge afhankelijkheden overstijgen meerdere bronnen van onzekerheid bij de planning en besluitvorming. Deze onzekere verandering genereert meerdere directe en indirecte effecten op SWS. Hierdoor is het moeilijk, en in sommige gevallen onmogelijk en onuitvoerbaar de omvang van de effecten op een systeem nauwkeurig te analyseren. De deterministische aanpak van een infrastructuur-systeemontwerp kan de kenmerken van een complex systeem en de onzekerheden die met toekomstige veranderingen samenhangen niet accuraat beschrijven. Daardoor is de kans zeer hoog dat niet aan het verwachte niveau van systeemprestaties kan worden voldaan en het systeem niet kosteneffectief is. Dit onderzoek gaat in op de belangrijkste wijzigingsdrukken voor de toekomst, gerelateerde onzekerheden en risico's voor duurzame infrastructuurplanning.

De meeste bestaande onzekerheidsmodelleringstechnieken zijn gebaseerd op ofwel kanstheorie ofwel op fuzzy verzamelingen-leer. De huidige economische problemen worden geassocieerd met de verschillende soorten van onzekerheid en in veel gevallen met dataschaarste. Een literatuuroverzicht voor onzekerheidsmodellering erkent dat een hybride aanpak van onzekerheidsmodellering nodig is om beide types van onzekerheid (dat wil zeggen, aleatorische en epistemische) te analyseren in een enkel raamwerk. Dit zou kunnen worden bereikt door (i) alle informatie in een probabilistische of fuzzy vorm te transformeren of homogeen te propageren middels een model (dat wil zeggen, hetzij willekeurige simulatie, hetzij fuzzy benadering), of (ii) alle informatie apart beschrijven en heterogeen te propageren. Er zijn echter weinig houvast om een onzekerheidsanalysetechniek te selecteren op

234

basis van hun bronnen en types. Bovendien beschrijven de meeste van de bestaande besluitvormingsraamwerken ontwikkeld voor op risico gebaseerde besluitvorming of multi-criteria analyses onzekerheid onvoldoende. Dit proefschrift ontwikkelt een onzekerheidsanalyseraamwerk en methoden die de meerdere bronnen en soorten van onzekerheid kunnen vangen in een enkel systeem.

Het besluitvormingsproces vereist het aanpakken van de tegenstrijdige sociale, economische en milieudoelstellingen. Dit kan worden bereikt met behulp van een *multi-criteria decision analysis* (MCDA). De gepubliceerde literatuur over MCDA toont aan dat berekende resultaten op basis van zeer eenvoudige technieken niet significant verschillen met resultaten van complexe methoden. Er werd gemeld dat MCDA methoden ofwel niet gebruiksvriendelijk zijn, ofwel niet transparant, of de betrokkenheid van belanghebbenden niet toelaten. Zo ontstond de behoefte aan een eenvoudige en transparante MCDA methode die de inzet van de belanghebbenden erkent. Bovendien is onzekerheid onvermijdelijk in het meten en het vertegenwoordigen van de prestatiecriteria. Daarom erkent de literatuur de noodzaak van een MCDA methode die onzekerheid kan vastleggen in een raamwerk, binnen de prestatiecriteria en met een techniek voor de aggregatie van meerdere informatiebronnen die kunnen variëren van scherp tot probabilistisch en kwalitatief.

Dit onderzoek heeft tot doel een besluitvormingsraamwerk en een methodologie te ontwikkelen voor het analyseren van de risico's en onzekerheden in SWS. Het onderzoek werd uitgevoerd door middel van theoretische en modelmatige benadering en demonstraties van het ontwikkelde raamwerk. Het heeft de volgende specifieke doelstellingen:

- Om een risicobeoordelingsraamwerk te ontwikkelen voor toekomstige veranderingsdruk in stedelijke watersystemen met de bijbehorende onzekerheden.
- Om een hybride benadering van de onzekerheidsanalyse te ontwikkelen die de verschillende soorten van onzekerheid in stedelijk watersysteemmodellen beschrijft en propageert.
- Om een multi-criteria besluitvormingsraamwerk te ontwikkelen dat is afgestemd op het runnen van een SWS onder onzekerheid.
- Om de toepasbaarheid en effectiviteit van de ontwikkelderaamwerken en methoden aan te tonen middels het toepassen op bestaande SWS.

De bestaande literatuur definieert risico niet concreet. Het wordt geïnterpreteerd als een combinatie van de waarschijnlijkheid van een negatieve gebeurtenis (b.v. wat er fout kan gaan) in een systeem en de bijbehorende gevolgen. Dit onderzoek definieert risico als een schending van het minimaal verwachte niveau van de prestaties en de bijbehorende negatieve gevolgen. Zo worden risico's voor SWS waargenomen wanneer de verwachte prestaties van het systeem worden geschonden.

Risico in een systeem wordt beoordeeld met behulp van een kwantitatieve of een kwalitatieve methode. Kwalitatieve methoden gebruiken deskundige adviezen om de kans op en de gevolgen van een mislukking te identificeren en evalueren. Vaak gebruikte kwalitatieve methoden omvatten het veiligheidsbeleidsreviews, checklists, "what if" scenario's, voorlopige risicoanalyse en risico en operabiliteitsstudies. In tegenstelling tot kwalitatieve methoden vertrouwen kwantitatieve methoden op informatiedatabases en statistische methoden. Een paar van de meest gebruikte kwantitatieve methoden van probabilistische risico-analyse omvatten faalwijzen en effectenanalyses, foutbomen en gebeurtenisbomen. Het geïdentificeerde risico in een systeem wordt verminderd of beheerst door het gebruik van verschillende maatregelen voorgesteld voor het risicobeheer. Risicomanagement is het proces van het selecteren van een strategie om het vastgestelde risico te beheersen of te verminderen. Gezien de complexiteit van SWS hanteert het voorgestelde risico-analyse-raamwerk zowel de probabiliteits- als de fuzzy set theorie om de kans op het risico van het falen van het systeem te analyseren en de MCDA methode om de beste strategie voor het risicobeheer te selecteren.

Dit onderzoek ontwikkelt een hybride methode voor onzekerheidsanalyse. In deze methode worden de invoerparameters van een model omgezet in een vorm (bijvoorbeeld in de probabilistische of fuzzy vorm) en vermenigvuldigd middels een homogeen model. De Dempster-Shafer theorie wordt gebruikt om de transformatie-algoritmen te ontwikkelen. De getransformeerde onzekere informatie kan worden verspreid met behulp van een gerenommeerde techniek, zoals Monte Carlo simulatie, Latin Hypercube sampling, Bootstrap simulatie en fuzzy alfa-cut technieken. De uitgang van de analyse wordt in een onzekerheidsenvelop met statistische gegevens en een 95% betrouwbaarheidsinterval.

Een op fuzzy set theorie gebaseerd multi-criteria besluitvormingsraamwerk wordt voorgesteld om de complexiteit en multi-disciplinaire aspecten van SWS aan te pakken. De multi-criteria zijn geselecteerd op verschillende dimensies van de infrastructuur (b.v. effectiviteit, betrouwbaarheid, systeemkosten) en meerdere doelstellingen van analyse (bijvoorbeeld milieu-, sociale, economische of andere doelstellingen). De criteria en indicatoren zijn vertegenwoordigd rekening houdend met de soorten informatie die verschillend verdeeld kunnen zijn. De samengestelde index wordt berekend met behulp van de fuzzy synthetische evaluatie techniek. Het relatieve belang van de indicatoren wordt bepaald door toepassing van een op een Analytisch Hiërarchisch Proces (AHP) gebaseerd algoritme. De index vertegenwoordigt de totale prestatie van een systeem of onderdelen van het systeem. De index wordt ook gebruikt voor het selecteren van een geschikte optie / strategie voor risicobeheer.

Dit onderzoek heeft de besluitvormingsraamwerken, methoden en instrumenten toegepast op drie verschillende echte gevallen. Het eerste geval analyseert de risico's verbonden aan de toekomstige beschikbaarheid van water in Birmingham, Verenigd Koninkrijk in 2035. De belangrijkste toekomstige veranderingsdruk beschouwd voor de beoordeling waren onder meer klimaatverandering,

236

bevolkingsgroei, sociaal-economische veranderingen, en waterverliezen uit het systeem. De hybride onzekerheidsanalysemethode wordt toegepast in deze case study. Het tweede geval is op basis van prestaties in de stedelijke infrastructuursysteemplanning in Kathmandu, Nepal. Het toont de toepasbaarheid van MCDA methode voor geïntegreerde besluitvorming voor de stedelijke infrastructuur. Het derde geval toont de stedelijke waterbronnen en analyse van de watervraag in een case situatie in Mbale, Oeganda met beperkte gegevens. Het laat zien hoe een op risico gebaseerd afwegingsraamwerk kan worden toegepast voor de SWS planning en besluitvorming.

Naast de kritische literatuurstudie over het definiëren van het risico-en onzekerheidsanalyseconcept, raamwerken en methoden voor onzekerheidsanalyse en multi-criteria besluitvorming zijn de belangrijkste bijdragen van dit onderzoek:

- Ontwikkeling van een nieuw risico-evaluatieraamwerk dat een hybride methode voorstelt voor onzekerheid, risico-evaluatie en een op fuzzy set theorie gebaseerde multi-criteria-analysemethode voor risicobeheer en de besluitvorming.
- Ontwikkeling van een hybride aanpak van onzekerheidsmodellering op basis van bewijstheorie om de verschillende soorten van onzekere informatie te analyseren in een enkel raamwerk.
- Ontwikkeling van een op fuzzy set theorie gebaseerd multi-criteria afwegingsraamwerk en een techniek voor besluitvorming op basis van prestaties en het selecteren van een duurzame strategie voor risicomanagement.
- Toepassing van de ontwikkelde raamwerken, methoden en tools voor SWS planning en besluitvorming in ontwikkelde en ontwikkelingslanden.

Het nieuwe raamwerk voor risicobeoordeling ontwikkeld op basis van dit onderzoek kan verschillende soorten en bronnen van onzekerheid beschrijven in een raamwerk en is ook van toepassing op andere terreinen van infrastructuurplanning zonder verdere wijziging . De hybride benadering van de onzekerheidsanalyse is de aangewezen richting voor de juiste vertegenwoordiging en het uitdragen van onzekerheden in een complex systeem. De MCDM raamwerk maakt een betere optie / strategie voor risicomanagement in SWS mogelijk. Het richt zich op multi-dimensionale aspecten van de infrastructuursysteemplanning en de rol van de belanghebbenden tijdens de besluitvorming. Het basis fuzzy set multi-criteria raamwerk voor de besluitvorming is flexibel genoeg om de verschillende soorten van gegevens die beschikbaar zijn in echte gevallen vast te leggen. De demonstratie van het ontwikkelde raamwerk, de technieken en tools in echte gevallen van SWS zowel in ontwikkelde als in ontwikkelingslanden toont de bruikbaarheid van de onderzoeksresultaten.

Krishna Bahadur Khatri

UNESCO-IHE, Delft

T - #0397 - 101024 - C252 - 240/165/14 - PB - 9781138000964 - Gloss Lamination